The Terry Lectures

THE SCIENTIFIC BUDDHA

Other Volumes in the Terry Lectures Series
Available from Yale University Press

The Courage to Be Paul Tillich
Psychoanalysis and Religion Erich Fromm
Becoming Gordon W. Allport
A Common Faith John Dewey
Education at the Crossroads Jacques Maritain
Psychology and Religion Carl G. Jung
Freud and Philosophy Paul Ricoeur
Freud and the Problem of God Hans Küng
Master Control Genes in Development and Evolution Walter J. Gehring
Belief in God in an Age of Science John Polkinghorne
Israelis and the Jewish Tradition David Hartman
The Empirical Stance Bas C. van Fraassen
One World: The Ethics of Globalization Peter Singer
Exorcism and Enlightenment H. C. Erik Midelfort
Reason, Faith, and Revolution: Reflections on the God Debate
    Terry Eagleton
Thinking in Circles: An Essay on Ring Composition Mary Douglas
The Religion and Science Debate: Why Does It Continue? Edited
    by Harold W. Attridge
*Natural Reflections: Human Cognition at the Nexus of Science and
    Religion* Barbara Herrnstein Smith
*Absence of Mind: The Dispelling of Inwardness from the Modern Myth of
    the Self* Marilynne Robinson
Islam, Science, and the Challenge of History Ahmad Dallal
*The New Universe and the Human Future: How a Shared Cosmology
    Could Transform the World* Nancy Ellen Abrams and Joel R.
    Primack

The Scientific Buddha

His Short and Happy Life

DONALD S. LOPEZ, JR.

Yale

UNIVERSITY PRESS

New Haven and London

Published with assistance from the Mary Cady Tew Memorial Fund.

Yale University Press books may be purchased in quantity for educational, business, or promotional use. For information, please e-mail sales.press@yale.edu (U.S. office) or sales@yaleup.co.uk (U.K. office).

Set in Janson type by Tseng Information Systems, Inc.
Printed in the United States of America.

Library of Congress Cataloging-in-Publication Data
Lopez, Donald S., Jr.
The scientific Buddha: his short and happy life / Donald S. Lopez, Jr.
    p.   cm. — (The Dwight Harrington Terry Foundation lectures on religion in the light of science and philosophy)
Includes bibliographical references and index.
ISBN 978-0-300-15912-7 (alk. paper)
1. Buddhism and science. I. Title.
BQ4570.S3L68 2012
294.3'365—dc23   2012015891

A catalogue record for this book is available from the British Library.

This paper meets the requirements of ANSI/NISO Z39.48-1992 (Permanence of Paper).

10 9 8 7 6 5 4 3 2 1

When we peruse the first histories of all nations, we are apt to imagine ourselves transported into some new world; where the whole frame of nature is disjointed, and every element performs its operations in a different manner, from what it does at present. Battles, revolutions, pestilence, famine and death, are never the effect of those natural causes, which we experience. Prodigies, omens, oracles, judgements, quite obscure the few natural events, that are intermingled with them. But as the former grow thinner every page, in proportion as we advance nearer the enlightened ages, we soon learn, that there is nothing mysterious or supernatural in the case, but that all proceeds from the usual propensity of mankind towards the marvellous, and that, though this inclination may at intervals receive a check from sense and learning, it can never be thoroughly extirpated from human nature.

David Hume, 1748

Contents

Preface   ix

Acknowledgments   xiii

ONE. A Purified Religion   1

TWO. The Birth of the Scientific Buddha   21

THREE. The Problem with Karma   47

INTERLUDE. A Primer on Buddhist Meditation   81

FOUR. The Death of the Scientific Buddha   101

Notes   133

Index   139

Preface

According to Buddhist doctrine, there have been many buddhas in the past, and there will be many buddhas in the future. We only remember one, called Gotama Buddha or Śākyamuni Buddha or, sometimes, "the historical Buddha." He lived in India twenty-five hundred years ago. Because he knew the past, he described the buddhas who had preceded him, some of whom he had encountered in his previous lives. Because he knew the future, he prophesied, by name, the buddha who would follow him.

Also according to Buddhist doctrine, there can be only one buddha for each historical age. A new buddha appears in the world only when the teachings of the previous buddha have been completely forgotten, with no remnant—a text, a statue, the ruins of a pagoda, or even a reference in a dictionary—remaining. Because the teachings of Gotama Buddha, Śākyamuni Buddha, the historical Buddha, that is, our Buddha, remain present in the world, it is said there is no need for a new buddha yet. According to the prophecies, the next buddha, whose name is Maitreya ("Benevolent"), will not appear in the world for millions of years.

But in the nineteenth century, a new buddha suddenly appeared in the world, a buddha unmentioned in any of the prophecies. I call him the Scientific Buddha. In the chapters that follow, I relate the story of his life and teachings, contrasting them with those of the older Buddha, the Buddha who appeared in India some two thousand five hundred years ago, the Buddha for whom the Scientific Buddha is commonly mistaken. My claim is that they are different, and that this case of mistaken identity has particular consequences.

This book derives from the Terry Lectures, more formally the Dwight Harrington Terry Foundation Lectures on Religion in the Light of Science and Philosophy, which I delivered at Yale University during the first week of October 2008. In the pages that follow, I have tried to retain at least some of the rhetorical flavor of those four lectures. The first chapter is presented largely as it was delivered, while the remaining chapters have been revised in various ways, with the last two chapters expanded considerably for this book. Chapter Three considers the doctrine of karma, the Buddhist Theory of Everything, and the ways in which it comes into conflict with the theory of evolution. The final chapter deals especially with the topic of meditation, a word that is used very widely today, and increasingly in the healthcare community. However, the traditional meaning of the term—a term that renders a number of different words in Buddhist languages—is not widely understood. For that reason, I have written a brief primer on Buddhist meditation, which is inserted between the third and fourth chapters.

In 2008, just at the time that I was delivering the Terry Lectures, I published a book titled *Buddhism and Science: A Guide for the Perplexed*. Because the present book is appearing just four years later, it is probably useful to say something about the difference between them. *Buddhism and Science* is largely a historical work, concerned above all with tracing the origins of the assertions for the compatibility of Buddhism and science, as well as with seeking to account for the uncanny persistence of those claims into the present. It does not, for the most part, assess the validity of such claims, nor does it offer a constructive alternative to them. Much of the historical material in this book, especially in the first chapter here, is drawn from *Buddhism and Science*, and readers interested in a fuller history can find it there.

The history of the discourse of Buddhism and science in the nineteenth and twentieth centuries offers a cautionary tale; many scientific elements proclaimed in the past to have been "anticipated" by the Buddha—from magnetism to radioactivity, auras, and a mechanistic universe—are claims that today appear as various versions of nonsense. With this history in mind, and with a cognizance of the structure of scientific revolutions, *The Scientific Buddha: His Short and Happy Life* examines two topics of considerable interest in the scientific (and pseudoscientific) communities, topics where the name of the Buddha is often invoked: evolution and meditation. Here, especially in the third and fourth chapters, I examine traditional Buddhist teachings on karma (often called a correlate to evolution) and meditation in an effort to determine the degree to which they conform to their popular representations. It will likely not give away the ending to say here at the outset that I find a significant dissonance between what karma and meditation have meant over the long history of Buddhist theory and practice in Asia, and how they are commonly understood today. I find that dissonance to be so deafening, in fact, that I suggest that we honor the Scientific Buddha for all he has done over his short life of 150 years (a mere trifle in the cosmology of Buddhism) and that we then allow him to pass away, like a flame going out.

Acknowledgments

I delivered the Terry Lectures at Yale University on October 1, 2, 6, and 7 in 2008. I would like to thank my hosts at Yale: Dale Martin, Phyllis Granoff, Koichi Shinohara, and Jacob Dalton of the Department of Religion, and Harry Attridge, Dean of the Divinity School. During my visit, I had the opportunity to meet with a number of impressive students, both undergraduates and graduate students. I was also welcomed by Jean Thomson Black and her colleagues at Yale University Press; it has been a pleasure to work with them on this project. Finally, I would like to thank Lauralee Field for all of her efforts in arranging the lectures and the many details before, during, and after my visit. That everything proceeded so seamlessly is testimony to the excellence of her work.

CHAPTER ONE

A Purified Religion

It is a great honor to be invited to deliver the Terry Lectures, and to be the first lecturer in their long history to address the topic of Buddhism. That history extends over more than a century, going back to Dwight Terry's original bequest in 1905, and the detailed instructions he provided for them in 1911. And the Terry Lectures are themselves part of a larger history of endowed lectures on religion.

These lectures on religion are each the products of their age, representing the concerns, and sometimes the prejudices, of their founders. Some of those concerns seem obscure and antiquated today; others remain relevant in their own way. In England in 1672, during the reign of Charles II, a group of Presbyterians and Independents established the Merchants' Lectures to "uphold the doctrines of the Reformation against the errors of Popery, Socinianism, and infidelity." The Socinians, largely forgotten today, called themselves the Polish Brethren and followed the teachings of Fausto Sozzini, who rejected the doctrines of the trinity and original sin. A century later, in 1768, the Bishop of Gloucester founded the Warburton Lectures, to "prove the truth of revealed religion in general, and of the Christian in particular, from the completion of the prophecies in the Old and New Testaments which relate to the Christian Church, especially to the apostasy of Papal Rome."

The late nineteenth century was something of a golden age for endowed lectures on religion. In a bequest in 1878, the Duff Lectures were established by the will of Alexander Duff, a Scottish missionary to India. That same year, at the Charter House of Westminster Abbey, Friedrich Max Müller, the leading Orientalist of the Victorian age, delivered the inaugural Hibbert Lectures, endowed by the Unitarian Robert Hibbert, who sought "the really capable and honest treatment of unsettled problems in theology." Perhaps the lectures most similar in spirit to the Terry Lectures were the Gifford Lectures, established in 1887 in the will of Adam Lord Gifford of Scotland. They were to be devoted to "Natural Religion, in the widest sense of that term," a topic that he wished to be "considered just as astronomy and chemistry is." His bequest stated that "the lecturer is to be subjected to no test at any time." By "test," he meant testimony, the requirement that one subscribe to a particular set of doctrines, such as the "Thirty-Nine Articles" of the Anglican Church. Indeed, he specified that the lecturers "may be of any denomination, or no denomination at all," as long as "they be able, reverent men, true thinkers, sincere lovers of and earnest inquirers after the truth."[1] The first Gifford Lectures were given by Andrew Lang, one of the founders of what would come to be called the Anthropology of Religion. In the United States, the Haskell Lectures, "on Middle Eastern Literature and its relation to the Bible and Christian teachings," were established by Caroline Haskell at Oberlin College in 1899.

The Terry Lectureship on Religion in the Light of Science and Philosophy, established at Yale University by Dwight Terry in

1905, was thus part of an important trend in Europe and America, a forum for reflection on the place of secular scholarship in Christian theology. Mr. Terry was not a Yale alumnus, or even a resident of New Haven. He was from Bridgeport, and his bequest specifies that "the lectures be repeated in his home city of Bridgeport, Conn., and elsewhere at the discretion of the managers." It is clear from his lengthy and eloquent bequest that he was a Christian, genuinely ecumenical in spirit, that he was convinced of the ultimate compatibility of religion and science, and of the benefits to be derived therefrom. We read: "The object of this Foundation is not the promotion of scientific investigation and discovery, but rather the assimilation and interpretation of that which has been or shall be hereafter discovered, and its application to human welfare, especially by the building of the truths of science and philosophy into the structure of a broadened and purified religion." He continues:

> The founder believes that such a religion will greatly stimulate intelligent effort for the improvement of human conditions and the advancement of the race in strength and excellence of character. To this end it is desired that lectures or a series of lectures be given by men eminent in their respective departments, on ethics, the history of civilization and religion, biblical research, all sciences and branches of knowledge which have an important bearing on the subject, all the great laws of nature, especially of evolution, . . . also such interpretations of literature and sociology as are in accord with the spirit of this Foundation, to the end that the Christian spirit may be nurtured in the fullest light of the world's knowledge and that mankind may be helped to attain its highest possible welfare and happiness upon this earth.

Like Lord Gifford, Mr. Terry insisted that the lecturers "shall be subject to no philosophical or religious test, and no one who is an earnest seeker after truth shall be excluded because his views seem radical or destructive of existing beliefs." Indeed, he noted that "the liberalism of one generation is often conservatism in the next, and that many an apostle of true liberty has suffered martyrdom at the hands of the orthodox." He thus welcomed "expressions of conviction from sincere thinkers of differing standpoints even when these may run counter to the generally accepted views of the day." He asked only "that the lecturers are well qualified for their work and are in harmony with the cardinal principles of this Foundation, which are loyalty to the truth, lead where it will, and devotion to human welfare."

If we examine Mr. Terry's bequest as a historical text, we see in it many of the marks of liberal American Protestantism at the turn of the century: an ecumenical spirit; a conviction that religion, or at least Christianity, is fully compatible with science; and that religion and science can work together, not so much to save the soul for the next life, but to benefit the mind and body in this one, improving the condition of humanity. There is still something of the missionary's zeal in his words, but the aim is not so much conversion to Christianity, but rather the use of science to nurture and perhaps revitalize what he calls "the Christian spirit" toward the larger goal of human happiness in this world; there is no mention of the next. In many ways, it is a sentiment typical of a particular optimism about religion, science, and the future of the human race in the years before the Great War. The perspective provided by the passage of a century might temper our own optimism, yet there is still much to admire in Mr. Terry's bequest.

If it were possible for me to speak with Mr. Terry today, I would suggest to him that there already exists a religion dedicated to helping mankind to "attain its highest possible welfare and happiness upon this earth," a religion that "will greatly stimulate intelligent effort for the improvement of human conditions and the advancement of the race in strength and excellence of character," that this religion proclaims a truth that both "has been" discovered and "shall be hereafter discovered," a religion that is also capable of nurturing the Christian spirit "in the fullest light of the world's knowledge." I would suggest to him that the truths of science and philosophy have already been built "into the structure of a broadened and purified religion." I would suggest to him that that purified religion is Buddhism.

Buddhism is a religion that has been described as both a philosophy and a science. It is a religion whose founder claimed to be neither a god nor a prophet of God, but a man, who took the title of Buddha, "the awakened one." This man, through his own efforts and his own investigations, discovered the most profound principles of the universe, and then compassionately taught them to others. In his first sermon, he taught what are known as the four noble truths: first, that life is qualified by suffering; second, that suffering has a cause; third, that there can be a permanent cessation to suffering; fourth, that there is a path to the cessation of suffering. This fourfold sequence reflects the scientific approach of the physician: the Buddha identified the symptoms, he made a diagnosis, he gave a prognosis, he prescribed a cure.

The Buddha described a universe that was not created by God but that functioned according to laws of causation. Indeed, the most famous statement in all of Buddhism is not a prayer, a mantra, or a profession of faith, but a summary of the Buddha's teaching: "Of those things that have causes, he has shown their causes. And he has also shown their cessation." This law of causation is not limited to the material world, but extends also to the moral realm, where virtue leads eventually to happiness and sin to suffering, not through the whims of a capricious God, but through the natural law of karma, a law that, unlike the theistic religions, offers an adequate answer to the question of why the innocent child must suffer.

The Buddha understood the operations of the mind in precise detail, explaining how desire, hatred, and ignorance motivate actions that eventually result in all manner of physical and mental pain, and he set forth the practice of meditation to bring the chattering mind and the unruly emotions under control in order to reach a state of serenity. But beyond this, he analyzed the myriad physical and mental constituents that together are called the person, finding among them nothing that lasts longer than an instant. Thus, he discovered, through his analysis, that there is no self, that there is no soul, that what we call the person is but a psychophysical process, and that the realization of this fundamental truth results in a certain liberation.

The Buddha then extended this analysis to the universe, declaring the universal truth of *pratītyasamutpāda*, dependent origination, according to which everything is interrelated, each entity connected to something, nothing standing alone, with effects depending on their causes, with wholes depending on their parts, and everything depending for its existence on the consciousness that perceives it. Yet, whether wave or particle, there is no uncertainty about the ultimate nature of reality, which the Buddha called *śūnyatā*, or emptiness.

The Buddha's discoveries were not limited to psychological truths and philosophical insights. He described multiple universes, each with its own sun, universes that arose out of nothingness and returned to nothingness over the course of vast cosmic phases of creation, abiding, and disintegration, phases measured in massive units of time called "countless eons." And he explained how countless beings are born in these universes during these countless eons, each moving, through a process of spiritual evolution, to a state of perfect wisdom.

The Buddha discovered these truths not through revelation but through investigation and analysis, testing hypotheses in the laboratory of his mind to arrive at proofs. He articulated these truths in his teachings, called the dharma, truths that derive not from faith, but from the Buddha's own experience. And having reached those conclusions, he did not declare them to be articles of faith, famously telling his followers: "O monks, like gold that is heated, cut, and rubbed, my words should be analyzed by the wise and then accepted; they should not do so out of reverence."

And when he died, he did not ascend into heaven. He lay down between two trees and said to his monks, "All conditioned things are subject to decay. Strive on with diligence." Then he passed away, like a flame going out.

I would tell all this to Mr. Terry, and for support I would note that in 1896, his contemporary Paul Carus, founder of the Open Court Press, had advocated what he called the Religion of Science, and had described the Buddha as "the first positivist, the first humanitarian, the first radical freethinker, the first iconoclast, and the first prophet of the Religion of Science."[2] Finally, I would point to the words of another of his contemporaries, Albert Einstein, who was developing the general theory of relativity at the time that Mr. Terry made his bequest. Here are Einstein's words: "The religion of the future will be a cosmic religion. It should transcend a personal God and avoid dogmas and theology. Covering both the natural and the spiritual, it should be based on a religious sense arising from the experience of all things, natural and spiritual as a meaningful unity. If there is any religion that would cope with modern scientific needs, it would be Buddhism."

This is Einstein's prophecy, a prophecy very much in the spirit of the Terry bequest. It declares that the religion of the future will not be limited to our single small planet but will encompass the entire cosmos. All religions make universal claims, but the implication here is that the religion of the future will do so accurately, its cosmology based not on fables but on physics. Endowed with a scientific insight into the nature of the cosmos, the religion of the future will be able to dispense with the primitive notion of a personal God who created the world, a God who bestows rewards and metes out punishments to his creatures. This religion will not bifurcate the spiritual and the natural but will reveal their harmony. It will require no tests, no creeds, no confessions of faith or assents to propositions that derive from the authority of some scripture, no rituals that reenact forgotten myths. It will have no dogma; it will have no church. There will be no damnation, only salvation, a salvation not through divine grace but through human experience, the experience of the individual's sense of oneness with the universe. And, finally, it will be compatible with science. This will be the religion of the future. But it is a religion that already exists. It is, indeed, an ancient religion. It is Buddhism, set forth by the Buddha over two millennia ago. Buddhism, I might again suggest, is the purified religion that Mr. Terry prophesied a century ago.

I might say all of that to Mr. Terry, but I would not believe it. Everything that I have just said about Buddhism has been said many times in the past. But it is only one perspective on a vast and ancient tradition, and it is a perspective that is both limited and limiting. For, everything that I have just said about the Buddha and Buddhism is either partial or misleading, despite the sense of comfort it creates. Indeed, even this commonly cited declaration of Albert Einstein appears nowhere in his writings or records of his conversations. It seems he never said it.

But there is something about Buddhism, and about the Buddha, that caused someone to ascribe this statement to Einstein, the Buddha of the Modern Age. And since the time when Einstein didn't say this, intimations of deep connections between Buddhism and science have continued, right up until today. On May 25, 2008, the Sunday *New York Times* published an article titled "Superhighway to Bliss" about Jill Bolte Taylor, a neuroscientist who suffered a massive stroke in 1996. After she regained the ability to speak, she described the experience as "nirvana." The next day, the piece was number one on the *Times* list of most e-mailed articles. In the "Science Times" section of the paper the following Tuesday, there was an article titled "Lotus Therapy," about the growing use of the meditation cushion to treat problems previously consigned to the analyst's couch. The next day, "Lotus Therapy" had taken over the top spot as the most e-mailed article. Two weeks earlier, the conservative commentator David Brooks titled his May 13 op-ed piece "The Neural Buddhists." The casual reader of the *New York Times* during the month of May in 2008 would likely have noticed this and wondered, "Why all the interest in Buddhism and science all of a sudden?"

Any scholar in the field of Buddhist studies would have been noticing this interest, at least out of the corner of the eye, for some years now. In my own case, I had originally imagined that claims for the compatibility of Buddhism and science derived from the 1960s, gaining their first popular expression in Fritjof Capra's 1975 bestseller *The Tao of Physics*. It turns out that I was both right and wrong. The claims did derive from the '60s; but I was off by a century. Statements about the compatibility of Buddhism and science were being made in the 1860s.

Identifying the historical origins of an assertion is the first step—a necessary but not a sufficient step—toward understanding that assertion, and it is significant that claims for the compatibility of Buddhism and science began in Europe and America during the Victorian period, as Buddhism became fashionable in intellectual circles. Similar claims for the compatibility of Buddhism and science began in Asia at the same time, a time when Buddhist thinkers were defending themselves against the attacks of Christian missionaries. Thus, to understand what the compatibility of Buddhism and science means today, it is necessary to understand what it meant a century and a half ago.

Buddhists first encountered science, perhaps ironically, in the guise of Christianity. When one reads missionary attacks on Buddhism, from Francis Xavier in Japan in the sixteenth century to Spence Hardy in Sri Lanka in the nineteenth century, Christianity is proclaimed as superior to Buddhism in part because it possesses the scientific knowledge to accurately describe the world, something that Buddhism lacked. For the missionaries, then, science was not an opponent of religion, or at least of the true religion, but its ally. Science would serve as a tool of the missionary and as a reason for conversion. Later, science would be portrayed as the product of a more generalized "European civilization," something that this civilization would take around the world; the vehicle for that journey was colonialism.

The efforts by Buddhist elites of the late nineteenth and early twentieth centuries to counter these claims and to argue that, on the contrary, Buddhism is the truly scientific religion (an argument that they seem to have eventually won) were directly precipitated by the Christian attacks. In a sense, the Buddhists wrested the weapon of science from the hands of the Christians and turned it against them. Whether to counter the missionary's charge that Buddhism was superstition and idolatry, or to counter the colonialist's claim that the Asian was prone to fanciful flights of mind and meaningless rituals of body, science proved the ideal weapon for the Buddhists. It was not Christianity but Buddhism, in fact, that was the scientific religion, the religion best suited for modernity, not just in Asia but throughout the world. Buddhism was the opposite of Christianity. Christianity has a creator God, and Buddhism has no God; Christianity has faith, Buddhism has reason; Christianity has dogma, Buddhism has philosophy; Christianity (at least certain kinds) has public ritual, Buddhism has private reflection; Christianity has sin, Buddhism has karma; Christianity has prayer, Buddhism has meditation; Christ is divine, the Buddha is human. And it was this human, this Asian, this Buddha, who knew millennia ago what the European was just beginning to discover. Some even went so far as to declare that Buddhism was not a religion at all, but was itself a science, a science of the mind. The implications of such a statement become evident in light of Victorian theories of social evolution, which saw the human race progressing from the state of primitive superstition, to religion, and then to science. As a science, Buddhism—condemned as a primitive superstition both by European and American missionaries and by Asian modernists—was able to leap from the bottom of the evolutionary scale to the top, bypassing the troublesome category of religion altogether.

A century and a half later, European missionaries have not disappeared, but their inroads into Buddhist societies are largely confined to specific times and places of the past: Japan in the sixteenth century, Sri Lanka in the early nineteenth century, Korea in the late nineteenth century. And European colonialism, in its classical form, has died out. Yet the discourse of Buddhism and science persists, unchanged in so many ways.

The leading Buddhist voices over the history of Buddhism and science have been disparate, coming from different Buddhist cultures, regarding different forms of Buddhism as the most authentic. Yet they also have something in common. Some were patriots in the struggle for Sri Lankan independence from Great Britain, others sought to prove the relevance of Buddhism to the new Republic of China, yet others argued for the essential role for Buddhism in the expanding empire of Meiji Japan. The inevitable links between nation and science help to explain why today the most famous proponent for the links between Buddhism and science is none other than the Dalai Lama, who has struggled for a half century for the independence of Tibet, perhaps still seeking to demonstrate that Tibetan Buddhism is not the primitive superstition of "Lamaism" that the European Orientalists condemned in the nineteenth century and that the Chinese Communist Party condemned in the twentieth. Rather, Tibetan Buddhism is presented as a worthy interlocutor of science and hence an appropriate ideology of a modern nation that might one day exist. The implication of politics in the discourse of Buddhism and science has been ironically clear in recent years, as the Dalai Lama has suggested that his successor might be selected through democratic election prior to his own demise, while the avowedly atheist Chinese Communist Party has countered that the traditional methods of divination must be followed to the letter.

But the question of Buddhism and science is not merely a historical question. Some may certainly choose to see it as such, and feel able to confidently consign it to the seemingly endless catalogue of products of European colonialism. However, others will not be satisfied with such an answer, feeling reluctant to see the links between Buddhism and science as anachronisms, seeking instead to know whether there might be some truth to the old claims, wondering whether there *are* ways in which Buddhism and science might in fact be compatible.

To ask that question raises a troubling thought: If Buddhism was compatible with the science of the nineteenth century, how can it also be compatible with the science of the twenty-first? If the Buddha long ago understood Newtonian physics, did he also understand quantum mechanics? Science has obviously made huge advances over the past century and a half in every domain, yet claims for the Buddha's prescience have remained persistent over this period. Furthermore, for the Buddhist, the content of the Buddha's enlightenment, however it has been defined, cannot change. It is an article of faith in all of the various forms of the tradition called Buddhism—whether it is a religion, a philosophy, or a science—that its truth can be traced directly back to the moment of the Buddha's enlightenment two and half millennia ago. How can the same timeless truths be constantly reflected in discoveries that have changed, and continue to change, so drastically over time?

Yet the claims for the compatibility of Buddhism and science have remained remarkably similar, both in their content and in their rhetorical form. This similarity has persisted despite major shifts in what is meant by "Buddhism," and what is meant by "science." In the early decades of this history, Buddhism generally referred to what European scholars dubbed "original Buddhism," the Buddhism of the Pali canon, preserved in the Theravada traditions of Southeast Asia and Sri Lanka. In the period after the Second World War, although the Theravada continued to be regarded as "Buddhism" in some quarters, Japanese Zen came to the fore. Since the 1990s, Tibetan Buddhism has displaced Zen to become the chief referent of Buddhism in the Buddhism and science dialogue, largely through the influence of the fourteenth Dalai Lama. And over the past decade, Buddhism has been identified with "mindfulness," a practice said to benefit the brain.

The referent of "science" is also fluid, sometimes nebulous. At times, science has meant a method of sober and rational investigation, with the claim that the Buddha made use of such a method to arrive at the knowledge of deep truths about the inner and outer world. At other times, science has referred to a specific theory: the mechanistic universe, the theory of evolution, the theory of relativity, the Big Bang, whose antecedents are to be found in the Buddha's teachings. At other times, science has referred to a specific technology—the microscope, the telescope, the spectrometer, and now the MRI—that has been used to discover what the Buddha knew long ago without the aid of such instruments; as more precise instruments have been developed over the past century, the claims of the Buddha's knowledge of their findings have remained constant. And at still other times, science has referred to the manipulation of matter, with dire consequences for humanity unless paired with the compassionate vision of the Buddha.

From the traditional perspective, the Buddhist truth is timeless; the Buddha understood the nature of reality fully at the moment of his enlightenment, and nothing beyond that reality has been discovered, or could be discovered, since that moment. From this perspective, then, the purpose of all Buddhist doctrine and practice that have developed over the past two and a half millennia is to make manifest the content of the Buddha's silent enlightenment, to express the inexpressible.

From the historical perspective, the content of the Buddha's enlightenment is irretrievable, and what is called "Buddhism" has developed in myriad forms across centuries and continents, with these forms linked by their retrospective gaze to that solitary sage seated beneath a tree. From either perspective—the traditional or the historical—in order to make this "Buddhism" compatible with "science," it would seem that the identity of the Buddha, and of what he taught, must be severely delimited, eliminating much of what has been deemed essential, whatever that might be, by the exalted monks and nuns and ordinary laypeople who have gone for refuge to the Buddha over the course of more than two thousand years.

If this is the cost, there must be a benefit. And this raises a question: Why do we yearn for the teachings of an itinerant mendicant in Iron Age India, even of such profound insight, to somehow anticipate the formulae of Einstein? I hope to be able to answer, or at least advance, this and other questions in the chapters that follow.

In those chapters, it will be necessary to use the word "Buddhism" many times. This is a term that scholars of Buddhism are generally reluctant to use in a declarative sentence, especially without an adjective or two. Indeed, in recent decades, scholars of Buddhism have even been reluctant to render that noun in the singular, speaking instead of "Buddhisms" in an effort to reflect the wide range of doctrine and practice across historical time and geographical space. Yet, I will use the word "Buddhism" many times in these chapters and will resist the urge to add a qualifying footnote in each case. I recognize that when I do so, some kind of counterexample is almost always available.

No scholar of Buddhism would dare attempt to identify some essence or even defining characteristic of Buddhism, instead offering, when asked, a rather dry historical narrative: "Buddhism is a religious tradition that began in India around the fifth century

B.C.E., founded by a figure known as the Buddha . . ." Although no scholar of Buddhism can say what Buddhism should be, a scholar can say, or at least speculate on the basis of historical evidence, what Buddhism has been for various groups of Buddhists across Asia, extending back over more than two millennia. And a scholar can say, or at least speculate on the basis of anthropological evidence, what Buddhism is for Buddhists, across Asia and elsewhere, during the present generation. It is clear that the Buddhism that is compatible with science must jettison much of what Buddhism has been, and is, in order to claim that compatibility.

For the Buddha to be identified as an ancient sage fully attuned to the findings of modern science, it was necessary that he first be transformed into a figure who differed in many ways from the Buddha who has been revered by Buddhists across Asia over the course of many centuries. During the Age of Discovery, as it was once called, European scholars divided the peoples of the world into four nations: Christians, Jews, Mahometans, and Idolaters. The Buddha was first encountered by European missionaries and travelers as but one of many idols, an idol known by many names. It was only in the late seventeenth century that the conclusion began to be drawn that the various statues seen in Siam, Cathay, Japan, and Ceylon, each with a different name, all represented the same god. And it was not until the early nineteenth century that it was known with certainty that that god had been a man, and that that man had been born in India. By that time, Buddhism was all but dead in India, and European scholars, many of whom never met a Buddhist or set foot in Asia, created a new Buddha, a Buddha made from manuscripts. This was the age of the quest for the historical Jesus. European philologists set out on their own quest for the historical Buddha, and they felt they found him. This Buddha was portrayed as a prince who had renounced

his throne, who proclaimed the truth to all who would listen, regardless of their social status, who prescribed a life dedicated to
morality, without the need for God. Such a savior held a special
appeal to Europeans and Americans in the last half of the nineteenth century, an appeal only heightened by the fact that unlike
Jesus, the Buddha was not a Jew, but an Aryan. It was this Buddha,
unknown in Asia until the nineteenth century, who would become
the Buddha we know today, and who would become the Scientific
Buddha.

Mr. Terry looked forward to the dawn of a new religion
forged from philosophy and science, a religion that will "greatly
stimulate intelligent effort for the improvement of human conditions and the advancement of the race in strength and excellence
of character." To this end, he asked that lectures be delivered on
"all sciences and branches of knowledge which have an important bearing on the subject, all the great laws of nature, especially of evolution." In the third chapter, then, in keeping with
Mr. Terry's request, I will focus on the theory of evolution and
its compatibility with the fundamental law of Buddhism, the law
of karma. In the first century of the Buddhism and science discourse, the parallels between the doctrine of karma and Charles
Darwin's theory were repeatedly proclaimed, both in Asia and in
the West, so much that karma came to be described by many as
a natural law. Indeed, it has often been claimed that Buddhism is
more scientific than the Abrahamic religions, in part because its
rejection of a creator god allowed it to avoid all the problems that
attend creationism. Only recently have Buddhist thinkers, most
notably the current Dalai Lama, argued that the theory of natural
selection has profound implications, in fact negative implications,
that strike at the foundation of Buddhist doctrine. I will examine
the classical doctrine of karma in some detail, I will evaluate its

importance to Buddhist thought, and then I will consider to what extent the Buddha is threatened by Darwin, focusing especially on Darwin's views on suffering. This chapter will include the postulation of a materialist Buddhism, one that does away with the mind and matter dualism that lies at the very heart of Buddhist thought. I will then consider: How much is lost in this process?

Having examined the past and, to some extent, the present, in the previous chapters, the last chapter will consider the future. It will ponder the question: If there can be fruitful conversations between Buddhism and science, how might they proceed? Even to pose such a question raises a host of issues, issues of what is meant by Buddhism, what is meant by science, and how the value of their merging might be measured. I will examine these issues in the context of the most famous form of Buddhist practice: meditation.

At the more recent turn of the century, meditation has become the centerpiece of the Buddhism and science discourse. Experiments are currently being conducted, data are currently being gathered, and that information is being broadly interpreted, with some scientists seeing more in it than others. My purpose here will not be to assess these findings, but to ask what these findings ultimately have to do with the meaning of meditation as it has been traditionally understood, and practiced, by Buddhists. That is, if forms of Buddhist meditation are shown to reduce what we today call "stress," what, if anything, does that mean? Is there a danger in turning Buddhism into a form of self-help, or has Buddhism always been, in its own way, a self-help movement?

Regardless of these cautions, the importance of meditation in Buddhism cannot be denied. But rather than merely seeking some form of clinical confirmation of Buddhist practice, whatever form that might take, it will also be important to use neurobiology to understand what happens in the meditative state. With this data in hand, we could then turn to another essential form of Buddhist practice, the practice of translation. Here, however, it would not be the translation of Sanskrit into Chinese, but the translation of the clinical description of Buddhist meditation into the traditional descriptions found in meditation manuals composed over the centuries. As a very tentative foray into this field, in the final chapter, I will use the findings of split-brain research to consider one of the most fundamental problems in Buddhist thought, the problem of the relation between doctrine and experience. That is, does doctrine generate meditative experience, or does meditative experience produce doctrine?

I will close with some final thoughts on why we ask that Buddhism be compatible with science and that science be compatible with Buddhism. I will ask what it would mean if we were to conclude that Buddhism is in fundamental ways incompatible with science. And I will ask whether that would really be so bad. Of all the things we have asked of the Buddha, must we also ask him to fulfill the dream of foreigners, a dream that has been dreamed in the West since at least the eighteenth century, the dream of something to believe in that does not make us have to choose between religion and science?

CHAPTER TWO

The Birth of the Scientific Buddha

When and where was the Scientific Buddha born? To answer this question, we begin with the story of the European encounter with the Buddha, and then go on to consider some elements of the Buddha's person that are not often mentioned in popular portrayals. My purpose is not to describe an original Buddha, and then a European deviation from it. Nor is it to suggest that there is a Western Buddha and an Eastern Buddha who are utterly unrelated. The situation is more complicated and more interesting.

European references to the Buddha go back to Clement of Alexandria in the third century C.E. In describing the Indian gymnosophists, Clement writes, "Some of the Indians obey the precepts of Boutta; whom, on account of his extraordinary sanctity, they have raised to divine honors." He goes on to say that the holy men of India "honor a kind of pyramid under which they believe the bones of some god are resting," a likely reference to Buddhist stupas, the domed structures (which turned into pagodas in China) that house the relics of the Buddha and his most famous disciples. Saint Jerome, writing in the fourth century C.E., reports that the gymnosophists of India believe that the founder of their religion, "Budda," was born from the side of a virgin. According to the traditional story, the future Buddha did indeed emerge from under his mother's right arm. But she was not a virgin, nor were Buddhist monks gymnosophists, that is, "naked philosophers," a term that would more accurately describe certain Jain and Hindu mendicants. After Clement of Alexandria, not much is heard about the Buddha in Greek or Latin until the eighth century, when, in the works of Saint John of Damascus, we find the story of Barlaam and Josaphat, a story that would become one of the most popular Christian saint's tales of the Middle Ages. It is clearly based on the life of the Buddha—the name of the protagonist, the prince Josaphat, derives from the Sanskrit term *bodhisattva*, or future buddha—although this would not be recognized until the nineteenth century, and in time to prevent his canonization.

References to the Buddha in European languages begin to appear again in the thirteenth century, largely due to European contact, often cordial contact, with the westward advancing Mongols. By 1259, the empire of Möngke Khan extended into Persia, Turkey, Russia, Georgia, and Armenia. In 1271, his brother Kublai Khan established the Yuan Dynasty, which would rule China for the next century. Returning to Venice after seventeen years in the service of Kublai Khan, Marco Polo's ship made port at the island of Sri Lanka, probably in 1292. He writes: "Furthermore you must know that in the Island of Seilan [Ceylon] there is an exceeding high mountain; it rises right up so steep and precipitous that no one could ascend it, were it not that they have taken and fixed to it several great and massive iron chains, so disposed that by help of these men are able to mount to the top. And I tell you they say that on this mountain is the sepulchre of Adam our first parent; at least that is what the Saracens say. But the Idolaters say that it is the sepulchre of SAGAMONI BORCAN, before whose time there were no idols. They hold him to have been the best of men, a great saint in fact, according to their fashion, and the first in whose name idols were made."[1]

"Sagamoni Borcan" is Mongolian for Śākyamuni Buddha, one of the most common names for the historical Buddha; *Śākyamuni* means "sage of the Śākya clan" in Sanskrit. The more important element of Marco Polo's account, however, is that he describes the Buddha as an idol and Buddhists as idolaters, labels that would remain in place for the next five hundred years.

As noted earlier, until long into the nineteenth century, European writers divided the world into four nations (in the sense of peoples): Christians, Jews, Mahometans, and Idolaters. What we think of as Buddhists were described by this last term, a word that we regard today, still feeling the power of Saint Thomas Aquinas, as a term of abuse. And in some sense, what we think of today as the academic study of religion has over the past two centuries been devoted to the process of separating out various groups of these idolaters and giving them their own religions, each ending in -ism: Hinduism, Sikhism, Daoism, Confucianism, Shintoism, Buddhism. Marco Polo, like the travelers who preceded and followed him, never identified the religion he encountered with the name "Buddhism" (or some rough Chinese or Mongol equivalent) or its priests as "Buddhists." Instead, he referred to the monks he encountered at the court of Kublai Khan and in his travels through Asia simply as idolaters, and the statues that they worshipped as idols.

From our perspective, it is clear that the statues those intrepid travelers described are not idols, and that what they term idols are statues not of different gods, but of a single man, the historical figure and founder of a great religion whom we know simply as the Buddha. But how could they have known that? Each of the Buddhist cultures of Asia visited by European travelers and missionaries—including those of China, Japan, Tibet, Sri Lanka, Thailand, Burma, and Vietnam—developed its own artistic conventions for representing the Buddha. Thus, not all images of the Buddha look alike. Furthermore, each of the Buddhist cultures of Asia had its own name for the Buddha, either translating or transliterating his Indian name, or one of his Indian epithets, into their own language. Those names and epithets were few; he tended to be known as either Buddha or Śākyamuni in China, Korea, Japan, and Tibet, and as either Gotama Buddha or Samaṇa Gotama ("the ascetic Gotama") in Sri Lanka and Southeast Asia. But in Chinese, Buddha is *fo;* the Chinese character pronounced in the eighteenth century as *fo* was, in the Tang Dynasty, pronounced *budh.* In Chinese, Śākyamuni is *Shijiamouni.* In Tibetan, Buddha is *Sangs rgyas* and Śākyamuni is *Shakya thub pa.* In Thailand, Samaṇa Gotama is *Samana Kodom.* These variations in local names for the Buddha were further variegated by how Europeans recorded those names that they heard. Francis Xavier reported that the Japanese worship an idol named Xaca. Matteo Ricci describes the "sect of Sciequia"; other Jesuits in China describe the idol Fôe. Ippolito Desideri, a Jesuit missionary to Tibet, reports that the lawgiver of the Tibetans is named Sciaccià-Thubbà. Robert Knox, a British sailor shipwrecked in Sri Lanka in 1659, reports that the Sinhalese worship Buddou. Simon de la Loubère, emissary of Louis XIV to the Court of Siam, relates the life story of Sommona-Codom. During the seventeenth and eighteenth centuries, just four names for the Buddha—Buddha, Śākyamuni, Gotama, and Samaṇa Gautama—would transmogrify into more than two hundred names in European accounts, including Boodhoo, Chekia-meouni, Godeman,

Sharmana Cardama, Sommonacodum, Tche-kia, and Xe Kian. By the end of the seventeenth century, the conclusion was beginning to be drawn that the various idols and their various names somehow represented the same figure, but whether he was mythological or historical, whether god, demon, or man, remained an open question.

The story of the European encounter with the Buddha is a long and fascinating one, too long to relate here. However, from the late seventeenth to the early nineteenth century a change, one might even say an evolution, occurred in the European perception of the Buddha, an evolution that eventually gave birth to the Scientific Buddha. In order to trace that evolution, let me provide four quotations, two from German scholars (or, more accurately, one Hessian and one Westphalian), one Scottish scholar, and one French scholar, beginning in 1667 and ending in 1844. Together with the changes in the Buddha they describe, we will also discern a change in those who describe him, moving from Catholic priest to amateur scholar to professional philologist.

The Catholic priest is Father Athanasius Kircher (1602–1680) of the Society of Jesus, one of the greatest scholars of the seventeenth century. Father Kircher never traveled to Asia. He composed the work from which the following passage is drawn, *China Illustrata*, based on reports and letters sent back to Rome by his fellow Jesuits. By 1667, the year of the publication of Kircher's Latin text, the Jesuits had been in East Asia for over a century, beginning with Francis Xavier's arrival in Japan in 1549 followed by Michele Ruggieri and Matteo Ricci's arrival in mainland China in 1583. Many Jesuit fathers would follow them, learning to read Chinese and Japanese well. Based on reports from the China mission, and his own research on the religions of India, Father Kircher provided the following account of the life of the Buddha:

The first creator and architect of the superstition was a very sinful brahmin imbued with Pythagoreanism. He was not content just to spread the doctrine, but even added to it so much that there is scarcely any one who is able to describe the doctrine or to write about it. He was an imposter known all over the East. The Indians called him Rama, the Chinese Xe Kian, the Japanese Xaca, and the Turks Chiaga. This deadly monster was born in central India in the place which the Chinese call Tien Truc Gnoc. His birth was portentous. They say his mother had a dream and saw a white elephant come first from her mouth and then from her left side. . . . So Xaca was born and he was the first who is said to have killed his mother. Then he pointed one hand toward heaven and the other down to the earth and said that except for him, there was none holy, not in heaven nor in earth. Then he betook himself to the mountain recesses and there he instituted this abominable idolatry with Satan's help. Afterwards he infected the whole Orient with his pestilent dogmas. The Chinese Annals say that when he emerged from his solitary hermitage, a divine (or more likely, a satanic) spirit filled him. He gathered together about 80,000 disciples. He selected 500 of these, and then 100 from these. Finally, he selected ten as being the best suited for teaching his horrible doctrines. He had chosen them as intimate counselors and associates in his crimes. Lest his doctrines be called in question by anyone, when dying, he decreed that the Pythagorean epithet be placed in his books. This phrase is, "He himself said," or, "So our books teach us." This means that it is evil to question the truth or the infallibility of these absurd fables, which are horrible and execrable. These are not tenets, but crimes. They are not doctrines, but abominations. They are not histories, but fables.[2]

This is obviously not a flattering portrait. Father Kircher begins by saying that the Buddha believed in Pythagoreanism. For the church, this did not mean that he was a student of geometry but that he believed in the transmigration of souls, or metempsychosis, a view that since antiquity had been attributed to Pythagoras, who is said to have heard the voice of a dead friend in the yelps of a puppy. For the church, the doctrine of reincarnation was a heresy. Father Kircher recognizes that the Buddha (a term he does not use) is known by a number of names, mistakenly adding that of the Hindu god Rama; according to later Hindu mythology, the Buddha was the ninth avatar or incarnation of the god Vishnu, with Rama being the seventh and Krishna the eighth. However, he correctly identifies India as the Buddha's place of birth, using the Vietnamese name (there were also Catholic missions in what is today Vietnam) for India, "Tien Truc Gnoc" (that is, Thien Truc Ngoc). Calling the Buddha by the Japanese name Xaca (that is, Shaka), he correctly notes that prior to his birth the Buddha's mother dreamed that a white elephant had entered her womb. And after he was born, he is said to have taken seven steps, pointed to the sky, and announced that he was the best of bipeds. The Buddha's mother died seven days after his birth, but nowhere does it say that he murdered her, as Father Kircher claims.

Four centuries before, Marco Polo had said that the Buddha was the first person to be honored as an idol. Here, Father Kircher describes the Buddha as instituting the practice of idolatry (with the help of Satan) and then spreading the practice all over Asia. It is his work, therefore, that the Christian missions must undo, made all the more difficult apparently by the fact that Buddhist texts begin with the phrase, "So our books teach us," presumably an allusion to the phrase that begins Buddhist sutras, "Thus did I hear" (*evaṃ mayā śrutam*). This demonic figure, purveyor of execrable fables, crimes, and abominations, is clearly not the Buddha that we know.

Just twenty-three years after the publication of *China Illustrata*, Engelbert Kaempfer (1651–1716) arrived in Japan. Kaempfer, a physician from Westphalia, had entered the service of King Charles XI of Sweden. Wishing to see the world, he joined a Swedish embassy to the court of Persia. Rather than return to Europe, he made his way to the Persian Gulf, where he boarded a

Dutch ship and signed on as Surgeon to the Fleet for the Dutch East India Company. His ship stopped in Arabia, the Malabar Coast in southwest India, then Sri Lanka, Bengal, and Sumatra, before reaching Batavia (Jakarta) in Indonesia, the headquarters of the Dutch East India Company, in September 1689. Eight months later, he was appointed as physician to the Dutch embassy to Japan. Sailing first to Siam, his ship arrived in Japan in September 1690. Dr. Kaempfer would spend the next two years there, most of it on Dejima, an artificial island in Nagasaki Bay to which Europeans were restricted. His history of Japan, written in High Dutch, was translated into English and published in London in 1727 as *The History of Japan, giving an Account of the ancient and present State and Government of that Empire; of its Temples, Palaces, Castles and other Buildings; of its Metals, Minerals, Trees, Plants, Animals, Birds and Fishes; of The Chronology and Succession of the Emperors, Ecclesiastical and Secular; of The Original Descent, Religions, Customs, and Manufactures of the Natives, and of their Trade and Commerce with the Dutch and Chinese. Together with a Description of the Kingdom of Siam.* The book contains a great deal of information about Buddhism, information that would be repeated in other works over the course of the eighteenth century. Having seen statues of the local god in both Thailand and Japan, Dr. Kaempfer offered the following account of their origin:

> The Siamites represent the first Teacher of their Paganism in their Temples, in the figure of a Negro sitting, of a prodigious size, his hair curl'd, the skin black, but as it were out of respect gilt over, accompanied on each side by one of his chief Companions, as also before and around him by the rest of his Apostles and Disciples, all of the same colour and most in the same posture. They believe according to the Brahmans, that the Deity dwelt in him, which he prov'd by his Doctrine, Way of Life, and Revelation. . . . The Ceylanese call him *Budhum*, the Chinese and Japanese *Sacka*, or *Siaka*, or plainly *Fotoge*,

that is, the Idol, and with an honourable Epithet *Si Tsun*, the great Saint.

About his origine and native Country, I find the account of those Heathens do not agree. . . . But the Siamites, and other Nations lying further East, have whole Books full of the birth, life and miracles of this God *Prah*, or *Siaka*. I am at a loss how to reconcile these various and opposite accounts, which I have gather'd in the abovesaid Countries, unless by supposing, what I really think to be the true opinion, *viz.* that the Siamites and other Nations lying more Easterly have confounded a younger Teacher with *Budha'* and mistaken the former for the latter, which confusion of the Gods and their names is very frequent in the Histories of the Greeks and Egyptians; so that *Prah* or *Siaka*, is not the same with *Budha*, much less with *Ram*, or *Rama*, as he is call'd by Father Kircher in his *Sina Illustrata*, the latter having appear'd many hundred thousand years before, but that he was some new Impostor who set up but about five hundred years before Christ's nativity. Besides this, many circumstances make it probable, that the *Prah*, or *Siaka*, was no Asiatick, or Indian, but some Egyptian Priest of note, probably of *Memphis*, and a Moor, who with his Brethren being expell'd their native Country, brought the Egyptian Religion into the Indies, and propagated it there.[3]

Kaempfer correctly identifies the various statues as representations of the same figure. But he cannot extract consistent information from his Asian informants about either the place or the date of his birth, so he offers two hypotheses that would hold sway for the next century. We might refer to them as the "two Buddha theory" and the "African hypothesis." These are often ascribed to Sir William Jones, who presented them to the Royal Asiatic Society in Calcutta in 1786 and 1788. But they were set forth almost a century earlier by Kaempfer, who derived the African hypothesis not from texts he read but from the statues he saw, and what he, and many others, would see as their wooly hair.[4] Kaempfer, and Father Kircher before him, had identified Egypt as the original source of the idolatry found in Asia. They saw the Persian conquest of Egypt in 525 B.C.E. as a pivotal moment, with the destruction of idols and the slaughter of priests. They speculated that those priests who were able to escape traveled east, reaching India. There, one of the Egyptian priests, named Prah (that is, Phra, "Eminent One" in Thai) or Siaka (that is, Shijia, Śākya in Chinese), in order to win followers in India, claimed to be the god "Budha," who had lived thousands of years before. Dr. Kaempfer identifies a number of shared elements to prove the Egyptian origin of Indian idolatry, including the worship of cows and belief in the transmigration of souls. It is important to note that Dr. Kaempfer does not differentiate what would later come to be known as Hinduism and Buddhism, but rather sees a single Asian paganism. However, the strongest piece of evidence for the African origin of the Buddha is to be seen in the statues, with their curly or "crisped" hair, which, once noted, would be seen by Europeans for the next century, with some adding the presence of thick lips and a broad nose. As late as 1825, the illustrious French Sinologist Jean-Pierre Abel-Rémusat felt it necessary to publish an article titled "On Some Descriptive Epithets of Buddha Showing that Buddha Did Not Belong to the Black Race," where he disputed the hypothesis that the Buddha was African.[5] For our purposes, the salient point is that in 1700, the Buddha was regarded as a historical figure—although his place of birth was disputed—who taught a form of idolatry that had spread across Asia.

Let us move forward by a century and travel from the island of Dejima in Nagasaki Bay in Japan to caves on Elephanta Island in Bombay Harbor in India. By 1800, the British were well established in India. The Dutch East India Company, which had employed Dr. Kaempfer, and which had dominated European trade with Asia for almost two centuries, lost its charter in 1800. Much of its territory had been taken by the British. The British East

India Company ruled most of India; India would become a British colony in 1858.

   William Erskine (1773–1852) traveled to India in 1804, at the invitation of his fellow Scotsman Sir James Mackintosh, eventually securing a position as clerk to the small cause court in Bombay. An amateur scholar, he assisted Mackintosh in founding the Literary Society of Bombay in 1804. Erskine taught himself Persian and devoted great efforts to a translation of the memoirs of Babur (1483–1531), the first Mughal emperor of India. However, he also developed an interest in the Buddhist, Jain, and Hindu temples in the region, including a remarkable group of caves filled with Hindu and Buddhist statues on Elephanta Island. On November 2, 1813, he read a paper to the Literary Society of Bombay titled "Account of the Cave-Temple of Elephanta with a Plan of the Drawings of the Principal Figures," which includes the following:

> When the Brahmins are taxed with idolatry, they always excuse themselves, as has been already remarked, by alleging the necessity of making an impression on rude minds by means of some intelligible symbols, on which the ignorant may fix their thoughts, and to which they may look for reward and punishment.
>
>   As in many of their incarnations the gods are supposed to have appeared with several heads, with the heads of animals, with a number of hands, and other singularities; their images in the temples correctly represent all these peculiarities. . . .
>
>   The religion of the Bouddhists differs very greatly from that of the Brahmins; as in the latter, God is introduced everywhere,—in the former, he is introduced no where. The gods of the Brahmins pervade and animate all nature; the god of the Bouddhists, like the god of the Epicureans, remains in repose, quite unconcerned about human affairs, and therefore is not the object of worship. With them there is no intelligent divine being who judges of human actions as good or bad, and rewards or punishes that as such;—this indeed is practically the same as having no God. Good and ill, according to their creed, are however supposed to spring invariably from virtue and vice; there

being as they believe an inseparable and necessary connexion between virtue and prosperity, vice and misfortune. Yet, as the mind of man must have some object of confidence on which to rest its hopes and to which to direct its supplication and prayer, they teach that from time to time men of surpassing piety and self-denial have appeared on the earth, and from their singular worth have after death been transferred to a state of superior bliss; which state, however, they say that we can only intimate by describing it as an absence of all pain, as we can only define health as an absence of all disease. These saints or prophets, after reforming the world in their lifetime, and by their superior sanctity attaining the power of performing miracles, are still imagined after death to have certain powers of influencing us. It is these men transferred by death to bliss who are the object of Bouddhist worship. This worship assumes different forms in different countries, and is by some supposed to be more widely diffused than any other religion. In Siam it is chiefly paid to Godoma or Sommona-Codom: but it is worthy of remark, that wherever this form of religion prevails in its original state, the relics of these holy men or saints are the object of worship. The largest temples are often in the form of a pyramid or of the section of a globe, and are supposed to contain a tooth, hair, or other relic of a saint. The forms of these holy places have been adopted from the custom prevalent in these countries of depositing the ashes of the deceased under a pyramid or globular mound: the pyramids are often of great size, and on their summits are umbrellas which are frequently adorned with bells; sometimes this pyramid is gilded over. Other temples of nearly similar construction, but hollow within, contain images to which adoration is directed. The images of these saints have different attitudes, sometimes sitting cross-legged in a meditative posture, sometimes standing upright.

As all the ideas of this religion relate to men, and as no incarnations or transformations of superior beings are recorded, it is obvious that in their temples we can expect to find no unnatural images, no figures compounded of man and beast, no monsters with many hands or many heads.[6]

Despite the separation of a century, Engelbert Kaempfer and William Erskine are similar in several ways. Each is in the employ of the most powerful trading company of his day; each is an amateur scholar fascinated by the culture he has encountered as a result of his professional assignment; neither reads the canonical languages of Buddhism; both draw historical conclusions based upon what they see in statues. But there are also significant differences. For Erskine, the religion of India is no longer an undifferentiated paganism. There are Brahmins and Buddhists. And the term "idolatry" is applied only to the Brahmins, something that Erskine implies that the Brahmins themselves concede when they are "taxed with" (that is, accused of) idolatry. The god of the Buddhists—no longer Pythagorean but now Epicurean—is not an object of worship; indeed, the Buddhists have no God. Instead, there is the law of karma, although Erskine does not use the term. His central point, supported by the statues that he saw, is that the religion of the Brahmins differs from the religion of the Buddhists because the Buddhists worship not monsters, but men, men who have achieved a certain sanctity.

This is a crucial turning point in the story of the Scientific Buddha. For the compatibility of Buddhism and science, there must be a scientific Buddha. In order for there to be a scientific Buddha, the Buddha must be a man, and not a god. In order for there to be a scientific Buddha, the Buddha must not be an idol, or teach idolatry. The transformation of the Buddha from god to man, from stone to flesh, was made possible by a number of factors, one of which was the physical location of the European author writing about the Buddha. Engelbert Kaempfer, who described the Buddha as an idol and Buddhists as idolaters, was living in Japan, where Buddhism was a living element of Japanese society, where Kaempfer could witness the worship of the Buddha by Japanese monks and laypeople. William Erskine was writing from India, where Buddhism was long dead, a relic of the past.

The European travelers to Asia—from Marco Polo in the

thirteenth century to William Erskine in the nineteenth—encountered Buddhists all over Asia, except in India, the land of the Buddha's birth, the land of the Buddha's teaching. The Buddha had lived in northern India in the fifth century B.C.E.; he is said to have lived for eighty years and likely died around 400 B.C.E. What we call Buddhism had flourished, or at least survived in India for a millennium and a half, during which time Buddhism spread to Central Asia, China, Korea, Japan, Tibet, and Mongolia to the north; to Sri Lanka and Southeast Asia to the south. Yet when Vasco da Gama landed in Calicut in southwest India in 1498, Buddhism had all but disappeared from India. The reasons for its disappearance were complex (a series of invasions of northern India by Muslim armies is one of many causes), but its consequences were profound. For by the time that European scholars (notably those of the British East India Company), trained in South Asian languages, began a sustained study of the culture and history of India, what they would come to call "Buddhism" was an artifact. There were no Buddhists in India, although there were Buddhists almost everywhere else in Asia. Instead, India had the ruins of stupas and monasteries to be excavated, inscriptions on stone pillars to be deciphered, and cave temples filled with statues (like the ones at Elephanta) to be illuminated and charted. These were artifacts of the past. In the present, India had no Buddhists but was filled with Hindus, who could be observed worshipping a pantheon teeming with monstrous gods. Instead of Buddhist monks, there were Brahmin priests who could be accused of idolatry, and made to answer.

The serenity of the Buddha that William Erskine describes is the serenity of the tomb, cool and silent, with gods, and the worshippers of gods, nowhere to be found. Had Erskine seen statues of the Buddha in Tibet, not so far to the north, he would have seen statues of the Buddha dressed in brocade robes, with pious worshippers repeatedly casting themselves to the ground in prostrations, offering lamps filled with burning butter. And he would have seen in the same temple any number of gods with multiple heads and arms, some with the heads of animals. Thus, a particular historical presence and a particular historical absence—the presence of statues of the Buddha and the absence of Buddhists—were crucial factors in the gestation and genesis of the Scientific Buddha.

I promised four quotes. The first one was from 1667. The last one, a long one, I fear, is from 1844.

> It is into the milieu of a society so constituted that was born, in a family of *kṣatriyas*, that of the Śākyas of Kapilavastu, who claimed descent from the ancient solar race of India, a young prince who, renouncing the world at the age of twenty-nine, became a monk under the name of *Śākyamuni* or also *Śramaṇa Gautama*. His doctrine, which according to the sūtras was more moral than metaphysical, at least in its principle, rested upon an opinion accepted as a fact and upon a hope presented as a certitude. This opinion is that the visible world is in perpetual change; that death succeeds life and life death; that man, like all that surrounds him, revolves in the eternal circle of transmigration; that he successively passes through all forms of life from the most elementary to the most perfect; that the place he occupies on the vast scale of living beings depends upon the merit of the actions he performs in this world; and thus the virtuous man must, after this life, be reborn with a divine body, and the guilty with a body of the damned; that the rewards of heaven and the punishments of hell have only a limited duration, like everything in the world; that time exhausts the merit of virtuous actions as it effaces the faults of evil actions; and that the fatal law of change brings the god as well as the damned back to earth, in order to again put both to the test and make them pass through a new series of transformations. The hope that Śākyamuni brought to humanity was the possibility to escape from the law of transmigration, entering what he calls nirvāṇa, that is to say, annihilation. The definitive sign of this annihilation was death; but a precursory sign in this life announced the man predestined for

this supreme liberation; it was the possession of an unlimited science, which gave him a clear view of the world, as it is, that is to say, the knowledge of physical and moral laws; and in a word, it was the practice of the six transcendent perfections: that of alms-giving, morality, science, energy, patience, and charity. The authority upon which the monk of the Śākya race supported his teaching was entirely personal; it was formed of two elements, one real and the other ideal. The first was the consistency and the saintliness of his conduct, of which chastity, patience, and charity formed the principal features. The second was the claim he made to be Buddha, that is to say, enlightened, and as such to possess superhuman science and power. With his power, he performed miracles; with his science, he perceived, in a form clear and complete, the past and the future. Thereby, he could recount everything that each person had done in their previous existences; and so he asserted that an infinite number of beings had long ago attained like him, through the practice of the same virtues, the dignity of Buddha, before entering into complete annihilation. In the end, he presented himself to humanity as its savior, and he promised them that his death would not annihilate his doctrine; but that this doctrine would endure for a great number of centuries after him, and that when his salutary action ceased, there would come into the world a new Buddha, whom he announced by name and whom, before descending to earth, the legends say, he himself had crowned in heaven, with the title of future Buddha.[7]

Perhaps unusual in its eloquence, this description seems conventional to us in its content.

In April 1837, twenty-four Sanskrit manuscripts arrived in Paris, sent from Kathmandu by a young officer of the British East India Company, Brian Houghton Hodgson (1800–1894), who was serving as British Resident at the Court of Nepal. They were Buddhist sutras and tantras, long lost in India but preserved by the Newars, a Buddhist community in the Kathmandu Valley. The Société Asiatique instructed two young scholars, both named Eugène—Burnouf and Jacquet—to examine the texts. Unlike Father Kircher, Dr. Kaempfer, and Mr. Erskine, Eugène Burnouf

(1801–1852) could read Buddhist texts in the original. Indeed, he was one of the great Sanskrit scholars of his day. Burnouf began with the *Perfection of Wisdom in Eight Thousand Stanzas*, but he found it repetitive and boring. Burnouf kept reading, turning next to the *Lotus Sutra*. He wrote to Hodgson: "Though many things are still obscure to my eyes, I nevertheless comprehend the progression of the book, the mode of exposition of the author, and I have even already translated two chapters in their entirety, omitting nothing. These are two parables, not lacking in interest, but which are especially curious specimens of the manner in which the teaching of the Buddhists is imparted and of the discursive and very Socratic method of exposition."[8]

We should note that the European estimation of the Buddha has now progressed from Pythagoras to Epicurus to Socrates himself. As it turns out, Burnouf had been captivated by arguably the single most influential text in the history of Buddhism, without knowing anything of its importance in India, China, Korea, and Japan. Four years later, in a letter of October 28, 1841, he informed Hodgson that he had finished printing his translation of the *Lotus Sutra*, "but I would like to give an introduction to this bizarre work." This introduction, from which the passage above is drawn, would transform the Western understanding of Buddhism.

Burnouf delayed the publication of his translation of the *Lotus Sutra* because he felt that it would not be comprehensible to European readers without an introduction. That introduction grew to 647 pages. Or to be more precise, the book that we have today, whose title page reads "Tome Premier," represents what Burnouf envisioned as the first volume of that introduction. He intended at least one and perhaps as many as three more volumes of introduction before he published his translation of the *Lotus Sutra*.

Introduction à l'histoire du Buddhisme indien appeared in Paris in 1844. It would become the most influential work on Buddhism written during the nineteenth century, setting the course for the academic study of Buddhism far into the twentieth. Its influence extended well beyond France. It was read in America by Ralph Waldo Emerson and Henry David Thoreau. It was read in Germany by Friedrich Schelling, Arthur Schopenhauer, and Friedrich Nietzsche. Richard Wagner wrote, "Burnouf's *Introduction to the History of Indian Buddhism* interested me most among my books, and I found material in it for a dramatic poem, which has stayed in my mind ever since, though only vaguely sketched."[9] Wagner's Buddhist opera, *Die Sieger*, was unfortunately never completed.

One of the most consequential sentences in the *Introduction* is buried in a footnote about halfway through the volume, where Burnouf writes, "The present volume is dedicated in its entirety to put in relief the purely human character of Buddhism." His book would play a crucial role in demythologizing and humanizing the Buddha, portraying a compassionate man who preached to all who would listen, without dogma and ritual. Burnouf writes, "I speak here in particular of the Buddhism which appears to me to be the most ancient, the human Buddhism, if I dare call it that, which consists almost entirely in very simple rules of morality, and where it is enough to believe that the Buddha was a man who reached a degree of intelligence and of virtue that each must take as the exemplar for his life."[10]

The Buddha had been transformed from an idol into a man, indeed, into a philosopher, a philosopher who rejected the rituals and myths of the Brahmin priests, a philosopher who set forth an ethical system, open to all, regardless of class and caste, a system based on reason. This transformation of the Buddha was made possible by the science of philology, the ability of Europeans to turn their gaze away from statues and toward texts, to look away from the Buddhists who stood before them across Asia, to look down at the Buddhist texts on their desks in Europe, and then imagine who the Buddha must have been. This was the moment when the Scientific Buddha was born. There had to be a science of philology before there could be a scientific Buddha.

What we have then is a familiar, and perhaps fitting, tale. European travelers, first mostly missionaries, merchants, and diplomats, and later colonial officers, encountered Buddhism across Asia. But they could not read Buddhist texts and hence they harbored many misconceptions about the religion and its founder. Eventually, with the rise of the science of philology and of Oriental scholarship in the nineteenth century, the huge corpus of Buddhist texts began to be read, providing a wealth of new information. The myriad idols coalesced into a single figure, who then became a historical figure, a founder of a religion, and a superstition became a philosophy. This is what used to be described unequivocally as progress.

This Buddhism would be built largely from texts. Because there were no Buddhists living in India during the colonial period, Buddhism, and especially what would come to be called "original Buddhism" or "primitive Buddhism," became the domain of European and later American and then Japanese scholars. They would create a Buddha and a Buddhism unknown in Asia, one that may never have existed there before the late nineteenth century. What they believed they had found, however, was the historical Buddha. This is all quite clear in the first academic lecture on Buddhism to be delivered in the United States. It was presented by Edward Eldridge Salisbury—instructor of Sanskrit at Yale, recently returned from study with Burnouf in Paris—at the first annual meeting of the American Oriental Society on May 28, 1844. Addressing the question of the historicity of the Buddha, Salisbury enumerated two points: "1. That a plausible foundation of real individuality is discoverable in even the wildest fables which veneration for Buddha has invented; and that the most extravagant have originated out of India, while nearly all agree in making India his native land. 2. That the images of Buddha are not monstrous, but seem to portray real humanity, while those of the old Hindu deities . . . are absurdly inhuman."

The son of a Congregationalist minister, Salisbury, after graduating from Yale in 1832, set out with his new bride for Europe, studying Arabic and Sanskrit in Oxford, Berlin, and Paris. He returned to Yale, where he offered, without salary, the first courses in Sanskrit in the United States. His lack of compensation may have been justified in part by the fact that over twelve years he only had two students. But one of these, William Dwight Whitney, went on to become a great Sanskritist. Indeed, when Salisbury had taught Whitney everything he knew, Salisbury resigned his professorship, arranged for it to be offered to Whitney, and then provided his own funds to endow the chair. For Whitney, the dream of every graduate student came true.

Salisbury's Buddha was the human and humane Buddha that his teacher Burnouf had described, a Buddha who has remained largely unchanged in the European and American imagination since then. It was also a Buddha who came to be adopted by Buddhist elites—in Sri Lanka, in Japan, in China—who were desperately seeking to demonstrate, either to Christian missionaries or to their own skeptical governments, that Buddhism was not a superstition, that Buddhism was a science.

The nineteenth century thus saw the birth of what the Jesuit historian Henri de Lubac has called the "scientific discovery" of Buddhism, that is, the study of Buddhism by scholars who could read Buddhist texts in the original. This was made possible by sig-nificant advances in the science of philology, with the discovery of language families and ancient connections between the classical Indian language of Sanskrit and the classical European languages of Greek and Latin, as well as modern German, French, and En-glish. These were called the Indo-European or Aryan languages; *āryan* is a Sanskrit term meaning "noble" or "superior." Through a complicated process, theories of language groups gave rise to theories of racial groups, and the kinship between the people of ancient India and the people of ancient Greece and hence (through a certain leap of faith) those of modern Europe became a matter not simply of verb roots but of blood lines. From this perspective, the Buddha was not so foreign. He was, in fact, racially an Aryan. But the nobility of the prince who had renounced his kingdom was not only hereditary. The Buddha had famously rejected the idea of inherited nobility, claiming that nobility derived instead from wisdom. He thus called his first teaching the four truths for the noble (rather than "the four noble truths," as the phrase has been famously mistranslated). In this way, the Buddha became doubly noble (*āryan*). He was noble by birth, blood, and language, yet he was also noble because he renounced his royal birth to achieve a spiritual nobility. In a Europe obsessed with questions of race and questions of humanity, the Buddha was both racially superior and a savior for all humanity, an ancient kinsman, a modern hero. This was a Buddha who would become a magnet for all manner of asso-ciations with modern theories over the course of more than a cen-tury, a Buddha to whom all manner of scientific insights would be ascribed, from the mechanisms of the universe to the structure of the atom, from a natural law of morality to the deepest workings of the mind.

To what extent is such a portrayal at odds with what we might term more traditional descriptions of the Buddha? Texts that we would classify as biographies of the Buddha did not begin to appear until several centuries after his death. The first of the Pali biographies, from the tradition that Victorian scholars thought most closely represented "original Buddhism," dates from the fifth century, almost a millennium after the Buddha's passage into nirvana. Buddhist texts are generally less interested in the events of the Buddha's life and more interested in his particular qualities, the qualities of his body, of his speech, and of his mind. His body is adorned with the thirty-two marks of a superman, including forty perfect teeth, a tongue long enough to lick behind his ears, arms long enough to rub his knees without bending forward, a protrusion on the top of his head, over which gods may not fly.

Many texts report that the Buddha smiled on occasion, but it seemed to be a tight-lipped smile, the subtle and often enigmatic smile we see in so many statues and paintings. It is noteworthy that we rarely see the Buddha grin, that is, smile in such a way that one could see his perfect teeth, forty teeth, teeth that are even, with no gaps between them, very white. There is an important reason for this. For when the Buddha smiles, it is said, multicolored rays of light radiate from his mouth and spread to the farthest reaches of the universe. They descend to cool the hot hells and to warm the cold hells. They ascend to the highest heavens, where the light is transformed into the sound of inspiring stanzas. The rays of light then return and disappear into the Buddha's body. The precise place at which they disappear is significant. If the Buddha wishes to explain a past action, the rays disappear into his back. If he wishes to predict the future, they disappear into his chest. If he wishes to predict a birth in hell, they disappear under the soles of his feet; if it is a birth among the animals, they disappear into his heel; if it is a birth among humans, they disappear into his knee;

if it is a birth among the gods, they disappear into his navel. If he wishes to predict that someone will achieve buddhahood, the rays of light disappear into the protrusion that crowns his head.

The Buddha's speech is endowed with sixty-four qualities of euphony. And if one thousand people simultaneously ask the Buddha a question in a thousand different languages, the Buddha understands each question and provides an answer in the questioner's language, although in fact he only utters the syllable *a*, the first syllable of the Sanskrit alphabet.

And what are the qualities of the Buddha's mind? He has complete recollection of the past, including each of his own past lives as well as the lives of all sentient beings (the past lives of each being are said to be infinite in number). He has full knowledge of the present in the sense that he is aware, or has the capacity to be aware, of all events occurring in all realms of multiple universes. He has full knowledge of everything that will occur in the future and is able to predict the precise circumstances under which various persons will become buddhas. He has direct and complete insight into the most profound nature of reality at all times, such that he does not need to enter into meditation in order to see reality directly. He is fully aware of the contents of the minds of all beings in the universe, and thus is able to teach them in accordance with their individual needs and dispositions. He is perfectly enlightened and hence omniscient. Although he passed into nirvana at the age of eighty, he could have lived for an eon or until the end of the eon, if he had only been asked. The Buddha described in the sutras is more a figure of science fiction than of science.

Scholars have long held that the buddhas who preceded Śākyamuni (typically counted as three or six or twenty-four) were not historical figures; each is separated from his predecessor by

billions of years. Instead, scholars speculate that the previous buddhas were introduced into the early tradition to counter charges
that the Buddha was guilty of the crime of innovation. The Buddha's opponents in the Brahmanical traditions of ancient India
claimed that their sacred texts, the Vedas, were eternal. They were
never composed but had always existed in the form of sound,
having been heard by the ancient sages. The existence of the previous buddhas allowed the Buddhists to make a similar claim. For
before our world system came into existence, there were previous
worlds with their own buddhas, without beginning. Each taught
the same truth, the same path to liberation from suffering. It is
the continuity with the buddhas of the past, rather than his unique
insight, that provided the foundation for the Buddha's authority.
Indeed, all buddhas are said to be remarkably similar in word and
deed. They differ from each other in only a few ways, one of which
is the circumference of their auras.

The Scientific Buddha is quite different. Here, the Buddha
is seen as a scientist, experimenting in his laboratory of the spirit,
trying first the life of indulgence and then the life of asceticism,
testing the various meditative techniques of the day before sitting
down under the tree on that full moon night and discovering the
truth for himself. The story is sometimes told in this way, and
perhaps this is what indeed occurred at that unrecoverable moment, subject to such endless commentary. But whether he was
the thousandth or the fourth or the first to see that truth, once it
was seen, there has been no further truth to be discovered over
the past two and a half millennia. To search for the truth in Buddhism is to follow the path that the Buddha revealed to the world.
There is no other path. In this sense, Buddhism is a profoundly
conservative tradition, constantly decrying innovation as deviation. Innovation has, of course, occurred in myriad ways over the

course of the tradition, but that innovation must always be portrayed as elaboration, as yet another articulation of the Buddha's silent enlightenment.

Yet, there is a certain parallel between the old Buddha of the tradition and the new Buddha of science. The Buddha of the tradition is validated by being the last, or more accurately, the most recent in a long line of enlightened beings who have discovered, and taught, the same truth. The Scientific Buddha is validated not by being at the end but by being at the beginning, as the perfected person who first discovered truths that lesser men would only learn millennia later. For the Buddha of the tradition to be valid, he must have understood what others had known long before him. For the Scientific Buddha to be valid, he must have understood what others did not know, and would not know, until long after him.

Each of these visions is profoundly retrospective, each evinces a deep longing for the primordial. The authority of the Buddha of the tradition derives from the fact that he has simply rediscovered eternal truths that the prehistoric buddhas had also found; much of the early literature recounts their lives in more detail than they do his. And the disciples of the Scientific Buddha derive deep comfort from the thought that the most modern discoveries, indeed, truths yet undiscovered, were known by this ancient Buddha, so long ago.

In closing, we might speculate as to what the traditional Buddha might say about the Scientific Buddha. In the *Great Discourse on the Lion's Roar* (Mahāsīhanāda Sutta)—a famous text from the Pali canon, the oldest collection of texts and, according to many, the collection that best represents his teachings—the Buddha declares, "Should anyone say of me: 'The recluse Gotama does not have any superhuman states, any distinction in knowledge and vision worthy of the noble ones. The recluse Gotama teaches a Dhamma [merely] hammered out by reasoning, following his own line of inquiry as it occurs to him'—unless he abandons that assertion and that state of mind and relinquishes that view, then as [surely as if he has been] carried off and put there, he will wind up in hell."[11]

These first two chapters have been historical and retrospective, exploring when, where, and how the claims for the compatibility of Buddhism and science came to be made. The next two chapters will be prospective and constructive, in the theological sense.

CHAPTER THREE

The Problem with Karma

In his bequest, Dwight Terry asked that lectures be given "on ethics, the history of civilization and religion, biblical research, all sciences and branches of knowledge which have an important bearing on the subject, all the great laws of nature, especially of evolution." The last term in his list should not surprise us, for in the debates about the relation between religion and science at the beginning of the twentieth century, evolution was clearly a charged topic. And one hundred years later, the power of Darwin's theory remains strong, so strong that, at least in the United States, all manner of state and local governments argue, often heatedly, about whether evolution should be taught to our children.

In 1849, Alfred, Lord Tennyson published "In Memoriam," a work he had originally entitled "The Way of the Soul." It is a long poem of 133 cantos ranging over many themes. About halfway through, there is a famous lament. In the fifty-sixth canto, Nature speaks:

> From scarped cliff and quarried stone
>   She cries, 'A thousand types are gone:
> I care for nothing, all shall go.

'Thou makest thine appeal to me:
  I bring to life, I bring to death:
  The spirit does but mean the breath:
I know no more.' And he, shall he,

Man, her last work, who seem'd so fair,
  Such splendid purpose in his eyes,
  Who roll'd the psalm to wintry skies,
Who built him fanes of fruitless prayer,

Who trusted God was love indeed
  And love Creation's final law—
  Tho' Nature, red in tooth and claw
With ravine, shriek'd against his creed

Who loved, who suffer'd countless ills,
  Who battled for the True, the Just,
  Be blown about the desert dust,
Or seal'd within the iron hills?

This section of "In Memoriam" was prompted not by Darwin (*On the Origin of Species* would not appear for another decade), but from Tennyson's reading of a work called *Vestiges of the Natural History of Creation*, a controversial work published anonymously in 1844; the author was later identified as Robert Chambers, a Scottish journalist. Yet, Tennyson expressed a fear that has persisted from the Victorian age and into our own, a fear that we are but passing elements of a material world, that we are but future fossils, our hopes and aspirations disappearing at our death, our virtue in vain, our prayers unanswered. Put into more modern language, we are dying machines designed for a single purpose: the survival of selfish genes.

Since the nineteenth century, the Buddha has been called upon to serve as an alternative savior, and Buddhism as an alternative religion, the religion that is not at odds with science, indeed the one religion that is compatible with science. That claim has been extended to specific scientific theories, and to the most challenging theory of all for religion, the theory of evolution.

Religions are generally averse to randomness. Yet two fundamental doctrines of Buddhism, the doctrine of karma and the attendant doctrine of rebirth, were seen then, and continue to be seen today, as somehow analogous to the theory of evolution. Thus the leading British scholar of Buddhism of the Victorian period, Thomas W. Rhys Davids, stated in his Hibbert Lectures of 1881, "And the more thorough-going the Evolutionist, the more clear his vision of the long perspective of history, the greater will be his appreciation of the strangeness of the fact that a theory so far consistent with what he holds to be true should have been possible at all in so remote a past."[1] We read in the *Frankfurter Zeitung* of April 25, 1890: "In itself Buddhism is a profound and all-embracing doctrine, adapted particularly for our time, because it does not contradict science, but contains on the contrary, the germs of scientific truths. For example, Transformism or Darwinism is involved in Buddhism."[2] And Thomas Huxley, Darwin's Bulldog himself, wrote in 1894: "Like the doctrine of evolution itself, that of transmigration has its roots in the world of reality; and it may claim such support as the great argument from analogy is capable of supplying."[3]

Much of the credit for the widespread perception of a compatibility between Darwin's theory of evolution and the Buddhist doctrine of rebirth during this period should go to the Theosophists. The Theosophical Society, founded in 1875 during the heyday of Spiritualism, is largely dismissed today, but in the decades bracketing the turn of the century, it included many influential thinkers and artists, in both Europe and America. Some of the first claims for the compatibility of Buddhism and science came from its founders, the Russian émigré Helena Petrovna Blavatsky and the Civil War veteran Henry Steel Olcott. Madame Blavatsky set forth her own theory of rebirth in works like *The Secret Doctrine* (1888), according to which entities that she called "monads" evolve over millions of years in a process of ascent through increasingly subtle forms. Blavatsky and her followers would claim that this system of spiritual evolution had already been intimated in the ancient mystery religions, including Buddhism. The Theosophist Walter Evans-Wentz observed in his first book, *The Fairy-Faith in Celtic Countries*, published in 1911, that Darwin's error was that he never identified exactly what it is that evolves; it cannot be the physical body, since that dissolves at death. Evans-Wentz writes, "Darwin thus overlooked the essential factor in his whole doctrine; while the Druids and other ancients, wiser than we have been willing to admit, seem not only to have anticipated Darwin by thousands of years, but also to have surpassed him in setting up their doctrine of re-birth, which explains both the physical and psychical evolution of man."[4] Evans-Wentz would make this claim more forcefully in his second book, *The Tibetan Book of the Dead.*

Yet the association of Darwin and the Buddha is not merely a remnant of a century ago. It persists into the twenty-first century. In 2007, Elite Books (whose motto is "Finding powerful new ways of relating") published *Dismantling Discontent: Buddha's Way Through Darwin's World.* And in 2009, the bicentenary of Darwin's birth, we find this headline in the *Times* of London for February 15: "Charles Darwin May Have Been Inspired by Tibetan Buddhism." The article explains that while preparing his new edition of Darwin's *The Expression of the Emotions in Man and Animals* (1872), the eminent psychologist and interpreter of the human face Paul Ekman found parallels with Buddhism, especially on the topic of compassion. Ekman then read some passages from Darwin's book to the Dalai Lama, who confirmed the affinities. This led Ekman to speculate that Darwin may have been influenced by Tibetan Buddhism, perhaps through his close friend the botanist Joseph Dalton Hooker, who, Dr. Ekman notes, "traveled to Tibet in 1847 and corresponded regularly with Darwin."

This chapter addresses the problem of karma, seeking to determine to what extent, if any, it is compatible with the theory of natural selection. We will find, I think, that, like Christianity, Buddhism is not immune to the challenges of Darwin. And I will take some comfort in the fact that, at least on this single point, I am not alone in calling the compatibility of Buddhism and science into question; the current Dalai Lama has recently expressed his own concerns.

Let me begin with a brief overview of the doctrine of karma, which is, in a sense, Buddhist science, the closest thing that Buddhism has to a Theory of Everything. In the *Majjhima Nikāya*, the "Middle Length Discourses" of the Buddha, we find the following exchange: A young Brahmin asks the Buddha, "Master Gotama, what is the cause and condition why human beings are seen to be inferior and superior? For people are seen to be short-lived and long-lived, sickly and healthy, ugly and beautiful, uninfluential and influential, poor and wealthy, low-born and high-born, stupid and wise. What is the cause and condition, Master Gotama, why human beings are seen to be inferior and superior?" The Buddha replies: "Student, beings are owners of their actions, heirs of their actions; they originate from their actions, have their actions as their refuge. It is action that distinguishes beings as inferior and superior."[5]

The word translated here as "action" is, of course, *karma*, a Sanskrit term of pre-Buddhist origin. But it is largely through Buddhism that it has made its way into the European lexicon, becoming the most commonly used Buddhist term in the English language, more than "Buddha," more than "nirvana." In common parlance, karma functions as a rough synonym of fate or luck, often bad luck. Despite its increasingly clichéd usage in English, however, it is difficult to understate the importance of karma in Buddhist thought. According to Buddhist cosmology, when a new universe evolves in the vacuity of space, the entire process begins when a faint wind starts to blow in a clockwise direction, becoming stronger and stronger until a circle of water forms above it, which in turn comes to be surmounted by a circle of earth. An entire world system eventually forms over the course of many eons, and when it is complete, it is populated by beings. That faint wind which begins to blow is the wind of the karma of the beings who will be reborn in the world that will eventually take form. The cause of the world is karma.

In Buddhist theories of gestation, the semen of the father and the ovum of the mother are merely the physical base for the new life. Life is the consciousness that arrives from a previous rebirth to enter the drop of the combined fluids. That consciousness carries with it all of the seeds of all the actions done by that being over a past that has no beginning, and it is a specific set of deeds—and not the genetic inheritance of the parents—that then ripen as the body and mind of the fetus. As the Buddha said, "beings are heirs of their actions."

Not surprisingly, karma is the subject of a vast literature in Buddhism, ranging from edifying morality tales to technical scholastic expositions, where we learn that karmic seeds always ripen in accord with their cause, that is, that virtuous deeds always result in happiness, that sinful deeds always result in suffering. We learn that because a single cause can have many effects, the store of karmic seeds is always growing. We learn that if a deed is not done, its result will not be experienced, that is, every experience of pleasure and pain, whether physical or mental, is the effect of a past deed. We learn that karmic seeds never lose their potency, no matter how long ago the deed was done. We learn that it is not only bodily actions that produce future effects; words and thoughts are also regarded as actions, and they also produce effects in the future. We learn that the nature of the effect is determined to a large degree not by the deed itself, but by the intention (*cetanā*) that motivates it.

In the scholastic treatises, we learn that there are deeds whose effects will be felt in this lifetime, deeds whose effects will only be felt in the next lifetime, and deeds whose effects will not be felt for two lifetimes or more. Thus, a person who grows rich by stealing from others may live happily in this life but will surely suffer in the future; a person who lives a virtuous life, serving others, may undergo privation in this life but will prosper in the future. We learn that there are deeds that affect only the doer and there are deeds whose effects overflow to others. We learn that a single deed done in the past serves as the primary cause for an entire lifetime, with a virtuous deed leading to birth as a god or a human, and a negative deed leading to birth as an animal, a ghost, or a denizen of hell.

Within this general division into the virtuous and the negative, there are detailed divisions and subdivisions, especially on the sinful side. It is explained that for a single action to serve as the cause of an entire lifetime, it must be a "complete action," in which four conditions must be met, conditions that again demonstrate the importance of intention. For example, if you decide to injure someone and end up killing him, it is not a complete action of killing. If you decide to murder one person and end up accidentally murdering someone else, it is not a complete action of killing. If you decide to murder someone and commit the crime, but have a change of heart before your victim actually expires, it is not a complete action of killing; if the change of heart occurs after the person dies, the action is complete. Unless the action is complete, it does not have the power to determine an entire lifetime. A murder that is a complete action will result in a lifetime in hell. The so-called incomplete actions of virtue or sin determine the conditions of a specific lifetime, such as physical appearance, health, and whether one is rich or poor.

We learn that the weight of a given deed depends upon a range of factors, and perhaps most important is the being for or against whom the deed is committed. Five deeds are known to deliver their doer directly into hell in the next lifetime, regardless of what virtues have been accumulated in the past. Those deeds are killing one's father, killing one's mother, killing someone who has achieved nirvana, wounding the Buddha, and finally, a deed that to our eyes seems much less grave: causing dissension in the monastic community. We learn that a single past deed can have three types of effect: a primary effect, an effect consistent with the cause, and an environmental effect. Thus, in the case of sexual misconduct, the primary effect, depending on the gravity of the deed, will be rebirth as an animal, ghost, or denizen of hell. The effect consistent with the cause is that when again reborn as a human, one will have an unfaithful spouse. The environmental effect is that one's house will always be dirty.

We learn that the moment of death is particularly important in determining the next lifetime. According to classical doctrine, it is not the case, as is often imagined (and is sometimes depicted in paintings of the Buddhist hells) that at the time of death, good deeds and bad deeds are weighed on a scale, with the heavier weight of accumulated deeds determining whether the next rebirth will be happy or sorrowful. Instead, a single complete action, performed in this life or any previous lifetime, comes to the fore at the moment of death and serves as the primary cause of the entire next lifetime. Thus, the state of mind at the time of death is critical in Buddhism, with the recollection of a virtuous deed and an attitude of calm said to help bring about a happy rebirth, and the recollection of a negative deed and an attitude of fear and regret said to help bring about rebirth in the lower realms.

Karma provides the explanation for everything. The physical environment is the product of the shared karma of the beings who inhabit it. Feelings of pleasure and pain, both physical and mental, are the product of past karma. Karma explains why some are rich and some are poor, why some are healthy and some are sick, why some live long and some die young.

Karma, then, is part of the natural order, but only one part. A Pali text called the *Niyamadīpanī*, perhaps translated as the *Explanation of the Laws* or the *Explanation of the Cosmic Order*, sets forth five constituents of the natural order. The first might be called the manifest order (*utuniyama*): that fire is hot, that water is wet, that earth is solid, that wind is motile; that each universe follows the same fourfold sequence of creation, abiding, destruction, and vacuity; that the three seasons—three because we are speaking of India—always follow in the same sequence: winter, summer, and the rainy season. The second element of the cosmic order might literally be translated as the "seed order" or the "seed law" (*bījaniyama*), the natural order as it pertains to plants: that a grain of rice grows into a rice seedling, that sugarcane is sweet. The third element of the cosmic order is the *karmaniyama*, the law of karma: that virtuous deeds naturally result in feelings of pleasure in the future and negative deeds naturally result in feelings of pain in the future. It is noteworthy that morality is placed here in the middle of a list of the five elements that constitute the cosmic order, suggesting that Victorian enthusiasts of Buddhism were perhaps justified in championing karma as a "natural law."

The fourth element of the cosmic order is that of consciousness, *citta*. This order encompasses what we might call Buddhist epistemology, the ways in which moments of consciousness are produced in a causal sequence, the relation of consciousness to its object, and, in the case of sense experience, the role of the sense organs. But we also find described here various paranormal powers: the divine eye, that is, the ability to see things that are far away; the divine ear, the ability to hear things that are far away; telepathy, that is, the knowledge of others' thoughts; the ability to remember one's past lives; as well as all manner of magical powers, such as flying through the sky while seated in the lotus posture, diving into the earth, walking through walls, and touching the sun and moon with one's hands.

The fifth and final element of the Buddhist cosmic order is the order of the *dharma*, a famously untranslatable term that here means something like the order of things. This would include the general law of cause and effect, in both the physical and moral realms, and the law of impermanence: that things come into existence, abide for a moment, and then pass away. But this natural order also includes the various portents that attend the birth of a buddha in his last lifetime, like the way that a special radiance pervades the entire world and the way that the earth trembles.

This list of the five elements of the cosmic order represents, in summary form, what might be termed traditional Buddhist science. What are its salient features? It is a natural philosophy, one in which the natural order extends into the moral realm. It describes an order in which everything operates with a predictable regularity, whether it is the sequence of the seasons, the growth of a sprout from a seed, or the tremors that attend the birth of a buddha. It describes an order that mixes, without apparent differentiation, what we moderns would term the natural and the supernatural. It also does not make distinctions between nature and the human; the natural world and human society are mutually implicated in their very origins.

So we now have some sense of the role of karma in the description of the world. But the Buddha did not simply describe the world, that is, the cycle of birth and death called *saṃsāra*, or "the wandering." He described a path to escape from that world. The purpose of the Buddhist path, as classically described, is to put an end to rebirth and the sufferings that inevitably attend it because of all the negative deeds that each being has done over an infinite past. Again, karma is the key. Rebirth is caused by karma. Karma—deeds of body, speech, and mind—is motivated by desire and hatred. Desire and hatred are caused by ignorance. Thus, if ignorance stops, desire and hatred become impossible. Without desire and hatred, there can be no karma because there is nothing to motivate the deeds that bind beings in saṃsāra. Without karma, there is no rebirth. And the wisdom that puts an end to ignorance can be cultivated to a sufficient strength that it not only makes the creation of future karma impossible, but it incinerates the seeds of all past actions. For someone who has destroyed those seeds, when the potency of the complete action that caused the present lifetime ends, rebirth does not recur. This is called nirvana.

All of this is summarized in the classical Buddhist doctrine of the four noble truths, a doctrine whose repeated contemplation is highly recommended, whether to the practitioner, the scholar, or someone who simply wants to know something about Buddhism. According to the traditional account, after spending his youth as a prince in an opulent palace, at the age of twenty-nine, Prince Siddhārtha went forth in search of a state beyond birth and death.

After practicing various forms of asceticism for six years, he sat down under a tree to meditate. He meditated all night, and by dawn he had awakened to the truth; the word for awakened in Sanskrit is *buddha*. Over the course of the next forty-nine days, he contemplated what he had seen that night, feeling at first that it was too profound to be understood by others. He eventually concluded that there were some who might understand, and he set out to find five former companions, known to history as "the group of five," who at that time were staying in a Deer Park outside Banaras. When he reached them, he briefly explained that one should follow a middle way between the extremes of self-indulgence and self-mortification, and then he said, "Now this, monks, is the noble truth of suffering: birth is suffering, aging is suffering, illness is suffering, death is suffering; union with what is displeasing is suffering; separation from what is pleasing is suffering; not to get what one wants is suffering; in brief, the five aggregates subject to clinging are suffering."

Suffering, then, was the first of the famous four noble truths. He went on to set forth the other three: that there is a cause of suffering, which he identified in this first sermon as craving, although elsewhere he would identify it as ignorance; that there is a state of the permanent cessation of suffering; that there is a path to that cessation. Thus, there are four truths: suffering, origin, cessation, path. In order to understand what the Buddha meant by the first truth, suffering, it might be useful to briefly describe all four truths.

The first truth is that existence is qualified by suffering. Like many other Indian teachers of the day, the Buddha believed in rebirth, that beings are reborn in various states throughout the universe. In Buddhism, these states are generally enumerated as six: gods, demigods, humans, animals, ghosts, and denizens of hell.

Although the sufferings undergone in the hot and cold hells are greater than those experienced in the heavens of the gods, each abode is ultimately unsatisfactory in one way or another, and all are unsatisfactory in the sense that even the happiest experiences eventually come to an end and one is reborn elsewhere.

The second truth, the truth of origin, explains why this is the case. The Buddha asserted that the proximate cause of suffering is karma. He denied the existence of an eternal God who is creator and judge, holding instead that physical and mental experiences of pleasure and pain, and indeed the environment in which they occur, are the results of past deeds. Virtuous deeds done in the past result in feelings of pleasure in the future, a future that comes either in the lifetime in which the deed is done or in any future lifetime. Negative deeds done in the past result in feelings of pain in the future. Virtue and non-virtue are generally defined respectively as helping and harming others, but are also enumerated. Thus, killing, stealing, and sexual misconduct are physical non-virtues; lying, divisive speech, harsh speech, and senseless speech are verbal non-virtues; and covetousness, harmful intent, and wrong view are mental non-virtues. These negative deeds are in turn motivated by a variety of negative mental states, the most important of which are desire, hatred, and ignorance, sometimes referred to as the three poisons. Among these, ignorance is identified as the root cause of all suffering. Although the nature of this ignorance is among the most extensively examined questions in Buddhist thought, it is generally defined as the belief in self: that among the physical and mental constituents of the person, there is a permanent, partless, autonomous self that is the owner of mind and body and that moves from one lifetime to the next. This fundamental misconception lies at the root of all suffering.

The third truth, the truth of cessation, postulates the existence of a state in which suffering has been destroyed. The existence of such a state may be inferred from the process by which suffering is produced. That is, suffering is caused by negative deeds; negative deeds are motivated by desire or hatred; desire and hatred result from ignorance. If ignorance can be destroyed, there can be no desire and hatred; without desire and hatred, there can be no negative deeds; without negative deeds there can be no suffering. The Buddha called the state of the cessation of all suffering and rebirth nirvana.

The fourth and final truth is the truth of the path, the path to that cessation. The path is described in many ways in Buddhist literature; one of the more succinct is known as the three trainings: in ethics, in meditation, and in wisdom. Ethical deeds, defined as the control of body and speech, result in states of happiness within the cycle of rebirth. But such deeds cannot, of themselves, bring about liberation from rebirth. Meditation in this context refers to practices that focus the mind, overcoming the ordinary state of mental distraction to produce a state of one-pointed concentration on a given object. Such concentration results in states of mental bliss, both in this lifetime and in future lifetimes, but does not, in itself, bring about liberation from rebirth. For liberation, wisdom is required, defined in this case as insight into the nature of reality, not simply intellectually, but at a level of deep concentration. The precise content of this wisdom is again a topic of detailed discourse in Buddhism, but it is generally described as the understanding that there is no self to be found among the constituents of mind and body.

I would note that upon hearing the Buddha describe the four noble truths, in less detail than I have given here, the group of five became enlightened.

Suffering, then, is the first constituent of the Buddha's first and most famous teaching. It is therefore unsurprising that it becomes a subject of considerable exegesis, both in other discourses of the Buddha and in later Buddhist literature. And, like so many things in Buddhism, it becomes the subject of lists. There are the four sufferings: birth, aging, sickness, and death. There are the eight sufferings of humans: birth, aging, sickness, death, losing friends, making enemies, not finding what you want, and finding what you don't want.

But as we set out to compare the meaning of suffering in Buddhism and in Darwin, one other list is particularly important. This is a short list, the list of the three sufferings: the suffering of pain, the suffering of change, and the suffering of conditioning. Let me describe each of these in turn.

The suffering of pain is the most obvious and includes the other sufferings—the four and the eight—just enumerated. This first form of suffering is what we ordinarily consider suffering: physical and mental feelings of pain, the pain that results from a physical injury, the pain that results from the loss of a loved one.

The Buddha, rather counterintuitively, identifies the second form of suffering, the suffering of change, as feelings of pleasure, which, by their very nature, will eventually turn into pain. The claim here is that pleasure and pain are fundamentally different: that pain remains painful unless something is done to alleviate it, while pleasure will naturally turn into pain. The warmth of the sun on one's back will eventually begin to burn; the satisfaction of savoring a delicious meal will turn to nausea if one continues eating. Indeed, there is no source of pleasure that will not become painful. This ancient Buddhist truth has recently been acknowledged in a ubiquitous television advertisement for Viagra that ends with the statement, "To avoid permanent injury, seek immediate medical attention for an erection lasting more than four hours."

The third of the three forms of suffering is called the suffering of conditioning: our minds and bodies are so conditioned that we are always subject to suffering in the next moment. This is because our existence depends upon causes and conditions over which we have no control, that is, the fruition of our past karma. The suffering of conditioning is both the most insidious and the most subtle; in a powerful analogy, the Buddha explains that the suffering of conditioning is like a tiny wisp of wool. For the unenlightened, it is like a wisp of wool in the palm of one's hand, so light and unobtrusive that one would not even know it is there. For the enlightened, the suffering of conditioning is like a wisp of wool in one's eye; so painful and distracting that one can think of nothing but its removal.

The subtlety of this suffering recalls one of the standard etymologies of the term "noble truth," or *āryasatya* in Sanskrit. The truths themselves are not noble. Rather they are called noble truths because they are true for the spiritually noble, a technical term used to refer to those who have gained at least an initial vision of nirvana. For such persons, suffering is true, origin is true, cessation is true, the path is true. For others, that is, for ordinary worldlings, they are not true. We do not understand that pleasure is a form of pain. We do not understand that each moment of our existence is a subtle form of suffering because pain can appear at any moment, without warning. We are thus doomed to suffering because we do not understand its nature.

The distinguished Victorian botanist Joseph Hooker departed England for the Himalayas in November 1847, arriving the following April in Darjeeling, where he stayed with the noted ornithologist Brian Houghton Hodgson, by then the former British Resident at the Court of Nepal, the same Hodgson who had sent Buddhist Sanskrit manuscripts to Paris a decade before. During his three-year expedition, Hooker collected plant specimens in Bengal, Nepal, and Sikkim, and he wrote often to Darwin. In May 1849, Hooker and Dr. Archibald Campbell, the superintendent of Darjeeling, crossed into Tibet. However, they were very soon turned back by Tibetan troops (all Tibetan Buddhists), and when they returned to Sikkimese territory, they were arrested on orders of the *dewan*, or prime minister, himself a devout Tibetan Buddhist. This created a minor international incident; Hooker and Campbell were mistreated by their captors (who were also Tibetan Buddhists), and their release was not negotiated until December.

I recount these events simply to suggest that Hooker would have had little reason to be kindly disposed to Tibetan Buddhism. And it appears that he made no mention of it in his letters to Darwin. In his two-volume *Himalayan Journals*, published in 1854, Hooker describes statues of Boodh, as he calls them, but does not discuss Buddhist doctrine, except to say that "the Sikkim Rajah has arrived at great sanctity, and is all but prepared for that absorption into the essence of Boodh, which is the end and aim of all good Boodhists."[6] Hooker may have learned something of Buddhism from Brian Hodgson, who had done important work on Newar Buddhism in the 1820s, but Hodgson remained rather critical of Tibetan Buddhism. So, it is doubtful that Darwin was influenced by Buddhism through Hooker. Few Europeans were able to read Tibetan texts at that time, and the Tibetan texts known to Europeans were for the most part translations of Indian Buddhist scriptures, and not the Tibetan treatises where the nature of compassion, and the techniques for its cultivation, are set forth. Under these circumstances, it is unclear how Darwin could have learned of Tibetan Buddhist theories of compassion.

Dr. Ekman was quoted in the *Times* to say, "I am certainly not saying Darwin was a Buddhist, but his view on the nature of compassion is identical in almost the exact words to the view of Tibetan Buddhism." This, I think, is also debatable. Darwin does not use the word "compassion" in *The Expression of the Emotions*, but rather "sympathy," and there only briefly, in connection with how and why certain emotions excite the lachrymal glands. But we must return to sympathy later.

And what of Darwin? From one perspective, it seems preposterous to compare the views of a mendicant of Iron Age India with those of a naturalist of Gilded Age England. There would seem to be massive problems of incommensurability, across time, across space, across languages, across cultures. The Buddha knew nothing of Darwin, of course, unless we grant him the omniscience that the sutras extol, and Darwin likely knew little of the Buddha. His rare mentions of "Hindoos" do not promise particular admiration; in his autobiography, he writes, "is it credible that if God were now to make a revelation to the Hindoos, he would permit it to be connected with the belief in Vishnu, Siva, &c. . . . ? This appeared to me utterly incredible."[7] How then could the Buddha and Darwin have understood each other? Yet it is a sign of the stature of each that they have come in many ways to stand outside time and place, that the truths they espoused can somehow be placed side by side.

Darwin wrote little about religion, it seems, unless he was forced by others to do so. To try to discern his views, it seems prudent to turn, then, to the chapter entitled "Religion" in the two-volume *Life and Letters of Charles Darwin*. And there we find an extract from a letter of June 5, 1879, written to Nicolai Alexandro-vitch Mengden, a seventeen-year-old baron studying in Tartu in Estonia, who wrote to Darwin in German to inquire about the religious beliefs of the seventy-year-old master. He had already written Darwin once and, apparently unsatisfied with the answer, wrote again, and this time Darwin patiently replied at some length on the topic. In the course of explaining why he no longer finds the argument from design to be convincing, Darwin takes up the question of how "the beneficent arrangement of the world can be accounted for." Let me quote what he has to say about happiness and suffering:

> Some other considerations, moreover, lead to the belief that all sen-tient beings have been formed so as to enjoy, as a general rule, happi-ness. . . . [The physical and mental organs] have been formed so that their possessors may compete successfully with other beings, and thus increase in number. Now an animal may be led to pursue that course of action which is most beneficial to the species by suffering, such as pain, hunger, thirst, and fear; or by pleasure, as in eating, drinking, and in the propagation of the species, &c.; or by both means com-bined, as in the search for food. But pain or suffering of any kind, if long continued, causes depression and lessens the power of action, yet is well adapted to make a creature guard itself against any great or sudden evil. Pleasurable sensations, on the other hand, may be long continued without depressing effect; on the contrary, they stimulate the whole system to increased action. Hence it has come to pass that most or all sentient beings have been developed in such a manner, through natural selection, that pleasurable sensations serve as their habitual guides.[8]

Here, Darwin says, with his typically careful reasoning, what we might expect him to say. Sentient beings seek to propagate their species. Those activities that promote such propagation are accompanied by feelings of pleasure, such that the pursuit of plea-sure in general contributes to the propagation of the species. We think immediately of sexual pleasure in this regard, but we might think also of the pleasure derived from food and drink (the issue of alcoholic drink raises a number of questions in this regard), and the pleasure derived from being sheltered from the elements. Suffering certainly exists, but it also has an evolutionary purpose in many cases. The suffering of hunger and thirst drives the animal to find food and drink so that it might survive. The suffering of heat and cold drives the animal to find shelter so that it might survive. The suffering caused by fear of the predator causes the animal to run and hide so that it might survive. Each of these sufferings serves the greater good of the propagation of the species. As Darwin goes on to write, "Such suffering is quite compatible with the belief in Natural Selection, which is not perfect in its action, but tends only to render each species as successful as possible in the battle for life with other species, in wonderfully complex and changing circumstances."

On the question of whether there is more happiness or more suffering in the world, then, Darwin comes down on the side of happiness, again for scientific reasons. Prolonged suffering leads to depression and inactivity, while happiness, or the promise of it, leads to the active pursuit of those things that provide pleasure, and pleasurable sensations may persist without deleterious effect. The species that remain are in a sense testimony to their successful pursuit of happiness.[9]

For the Buddhist, the origins of life are not random. All sentient beings are alike in that they want happiness and do not want suffering. Yet their desires and their deeds are at cross-purposes. Seeking happiness, they engage in deeds motivated by the desire to satisfy a self that, in fact, is an illusion. Thus, although they strive for happiness, they accumulate the causes of more and more suffering, dooming themselves to wander forever in the cycle of birth, aging, sickness, and death. Suffering, then, is a noble truth only in the sense that the pervasiveness of suffering is recognized by the spiritually noble. Suffering itself is ignoble, the bitter fruit of karmic seeds sown by deeds motivated by desire, hatred, and ignorance in a world of uncontrollable change. According to a Buddhist saying, "Everything that is independent is happy, everything that is dependent suffers." This is not a political slogan but a metaphysical dictum. The term translated here as "dependent" is *paratantra* in Sanskrit, literally "under the control of another." According to the Buddha, all sentient beings are under the control of their own past deeds, and so they suffer. Happiness or the promise of it leads, as Darwin observes, to the active pursuit of those things that provide pleasure. But for the Buddhist, those pleasures are unsatisfying, for the objects of desire are all impermanent, disintegrating at every moment, like raindrops falling on a courtyard, like pottery smashing on a stone floor. The species that remain are, in a sense, testimony to their failed pursuit of happiness. They have forever sought happiness. They have forever found only suffering and perpetual rebirth.

In that same letter to Baron von Mengden, Darwin writes, "If all the individuals of any species were habitually to suffer to an extreme degree, they would neglect to propagate their kind; but we have no reason to believe that has ever, or at least often occurred."[10] In a certain sense, this is precisely what occurred in the case of the Buddha. By setting forth the four truths and by describing the six abodes of rebirth—each abode branded with the three marks of suffering, impermanence, and no self, each abode also with its own kind of pain—six abodes where each being has been reborn countless times in the past and will be reborn countless times in the future unless the antidote is found, the Buddha is suggesting that all individuals of all species have always suffered to an extreme degree. Furthermore, for those who have become convinced of this truth, he established an order of monks and nuns, designed as the ideal setting for those who seek a permanent release from all forms of suffering. One of the conditions of entry into the order, for both men and women, is that they vow to neglect to propagate their kind. It could be argued that this is just another example of pain motivating the pursuit of pleasure that Darwin describes. Buddhist monks and nuns suffer the pain that comes from renouncing sexual pleasure in order to experience a greater bliss, the bliss of nirvana. But it is difficult to imagine that the permanent cessation of the corporeal and the conscious called nirvana is something that Darwin would have included in his rather more quotidian category of pleasure.

There certainly might be ways in which we could find the Buddha's understanding of suffering to be somehow compatible with that of Charles Darwin. But I wonder whether, in doing so, we do not run the risk of demeaning them both, effacing the nuance from each in order to derive a faded facsimile. Still, we might take from Darwin the idea of a species that persists through generations and across time. And we might say that for the Buddha there are not myriad species, but a single species, the species of the sentient, to which all forms of sentient life belong, both those that Darwin would recognize (insects, fish, birds, mammals) and those that he would not (the gods in their heavens, the ghosts whose search for food is always frustrated, the denizens of hell). That instead of moving up the tree of life, the members of this species have wandered among its many branches forever, sometimes climbing up, sometimes falling down.

But what does this have to do with Darwin's theory of evolution? It would seem that the Buddhist doctrines of karma and rebirth correlate poorly with the theory of natural selection,

in which random mutations of matter, not mind, occur uncon-
sciously over long periods of time in such a way as to insure the
survival of a species from one generation to the next. And, indeed,
in his book *The Universe in a Single Atom,* the Dalai Lama writes,
"From the Buddhist perspective, the idea of these mutations being
purely random events is deeply unsatisfying for a theory that pur-
ports to explain the origin of life."[11] This is another way of saying
that natural selection is at odds with the doctrine of karma. And
the reason it is at odds can be traced to the one domain in which
Buddhists are strict dualists, and that is on the question of the relation of mind and matter.

It is a fundamental tenet of Buddhist philosophy that cause
and effect must be of the same substance, or *dravya;* a seed is the
same substance as the sprout it produces, a pot is the same substance as the clay from which it is made. Thus, mind and matter,
although closely related in many ways, are different substances,
such that mind cannot be produced from matter. As the Dalai
Lama states, "According to Buddhism, though consciousness
and matter can and do contribute toward the origination of each
other, one can never become the substantial cause of the other."[12]
Consciousness is the product of consciousness; mind comes from
mind. The present moment of consciousness is the product of a
previous moment of consciousness. The consciousness of a new-
born infant is the product of the consciousness of the fetus in the
womb. The consciousness in the womb is the product of the consciousness present at the moment of conception. The consciousness present at the moment of conception is the product of the
consciousness at the moment prior to conception. That moment
of consciousness can only come from a previous lifetime. This is
the classical Buddhist proof of rebirth.

Furthermore, in Buddhism, rebirth, and hence consciousness, have no beginning. The Buddha is said to have refused to answer questions about origins and endpoints; such concerns are counted among the fourteen questions, called the "unindicated views" (*avyākṛta*), to which the Buddha remained silent. The great physicist J. Robert Oppenheimer, himself a student of Eastern philosophy, saw a parallel here: "To what appeared to be the simplest questions, we will tend to give either no answer or an answer which will at first sight be reminiscent more of a strange catechism than of the straightforward affirmatives of physical science. If we ask, for instance, whether the position of the electron remains the same, we must say 'no'; if we ask whether the electron's position changes with time, we must say 'no'; if we ask whether the electron is at rest, we must say 'no'; if we ask whether it is in motion, we must say 'no.' The Buddha has given such answers when interrogated as to the conditions of a man's self after his death; but they are not familiar answers for the tradition of seventeenth- and eighteenth-century science."[13]

However, the Buddha's apparent equivocation here had a somewhat different purpose. He warned that refusing to follow the religious path until one understands the origin of the world is to be like the man who refused to have a poison arrow extracted from his body until he knew whether the person who shot the arrow was tall, short, or of medium height. The great Harvard Pali scholar of the nineteenth century, Henry Clarke Warren, called these "questions which tend not to edification." The Buddha described all such questions as "a jungle, a wilderness, a puppet-show, a writhing, and a fetter, and coupled with misery, ruin, despair, and agony." Thus, it is not that the answers to such questions cannot be known; it is that, under the dire circumstances of suffering and rebirth, they are not worth knowing.

This places the Buddhist thinker in a difficult position with regard to investigations of the origin and nature of consciousness. But it is clear that in Buddhism, consciousness cannot emerge from just anything, or from nothing. In Buddhism, consciousness simply has no beginning, with one moment of consciousness having been produced by the one before it, stretching back infinitely. This is not to say that change does not occur over time, but the change is a process of devolution. According to the standard Buddhist cosmology, the human race is in a process of decline: in height, in intelligence, in virtue, in lifespan. Indeed, the average human lifespan, now said to be one hundred years, will gradually diminish over the eons until it is only ten years, a degenerate age of constant warfare among fourth-graders (certainly the most horrifying Armageddon foretold in any religion), at which point the lifespan of humans will increase before the next cycle of decline. But from the perspective of evolutionary theory, the crucial point is that in Buddhism, because mind and matter are different substances, there can be no transition from the non-sentient to the sentient.

The theory of evolution explains how species mutate through natural selection and thereby survive. The aim is the perpetuation of the species, and this happens randomly, without intention. Yet from the Buddhist perspective, the entire notion of a species becomes problematic. One crucial consequence of the beginningless cycle of rebirth is that each sentient being has been a god, a demigod, a human, an animal, a ghost, and a denizen of hell many, many times in the past. Each being has been every species. In a certain sense, there is only one species, this species of the sentient, with phenotypic variation occurring not as a result of natural selection but as a result of karma, with that variation changing radically and, to the sight of all but the enlightened, randomly from lifetime to lifetime. The wheel of rebirth is not a process of evolution, even spiritual evolution, as many have wanted to believe, a process where beings are born as animals, then humans, then gods, before entering nirvana. Instead, the beings wander up and down throughout the six realms; gods fall into hell and monkeys go to heaven. But that randomness is only apparent. Change occurs because of specific karmic causes. A famous Buddhist text, the *Garland of Jewels* (Ratnāvalī) by the second-century C.E. master Nāgārjuna explains that the performance of specific ethical deeds in one lifetime will result in specific physical attributes in a future life.

> Through using true and non-divisive
> Speech over a long time
> You will have forty glorious teeth
> Set evenly and good.
>
> Through viewing beings with love
> And without desire, hatred, or delusion
> Your eyes will be bright and blue
> With eyelashes like a bull.[14]

Furthermore, in Buddhism, there is no dispersion of the gene pool from one generation to the next. Each new body in the cycle of rebirth receives the full inheritance of all the karma from all the past lives of that particular series of moments of consciousness. And, left unimpeded, that continuum of consciousness will continue forever. The goal of the Buddhist path is to stop it. The words that are traditionally used to describe this goal are cessation, extinction, nirvana.

Extinction in nature is the disappearance of a particular species, such that the reproduction of a subsequent generation becomes impossible. This may occur through a loss of habitat and the inability to move to new habitats, through predation, through inbreeding, through the introduction of non-native species, through pollution, through a catastrophic event, through disease. All species, it seems, will naturally become extinct, on average, after a period of about ten million years. The species that remain today represent only one tenth of one percent of the species that have existed. The other 99.9 percent are extinct. Still, the survival of species, through natural selection, remains the engine of existence.

In Buddhism, the highest form of creation is extinction. And, unlike extinction for Darwin, for Buddhism, this extinction cannot occur naturally; it can only occur as the result of conscious intervention. Here, the tropes of the tradition suggests that the species of the sentient is not endangered by pollution; the species of the sentient is preserved by pollution. Pollution, of course, does not infect the body; it infects the mind. Negative mental states, called the *kleśas* or "afflictions" in Buddhism, disrupt, muddle, and confuse the mind, displacing virtuous states and inducing present agitation and future suffering; the impurity that the afflictions discharge is not easily expunged, causing pain over the course of many lifetimes. And these defilements are organic to the extent that they are states of mind that occur quite commonly, we might even say, naturally: desire, anger, pride, doubt, jealousy, lethargy, agitation, laziness, forgetfulness, distraction. Permanent separation from these states is the goal of the Buddhist path. That goal, called nirvana, is not the suppression of these negative states of mind, but their eradication, an eradication which, at least in the early tradition, entails the permanent extinction of mind and body, an extinction that occurs because the causes of their reproduction have been destroyed. There can be no mind and body because there is no karma to cause them. There can be no karma because there is no desire or hatred to motivate deeds. There can be no desire and hatred, because ignorance, that is, the belief in self, has been destroyed. Among the many wisdoms of the Buddha, there is this famous pair: the knowledge that ignorance has been destroyed and the knowledge that it will never be produced again.

So there seems to be a certain parallel between the survival of the species and the perpetuation of the cycle of rebirth. As long as sentient beings—the gods, demigods, humans, animals, ghosts, and denizens of hell who populate the realms of *saṃsāra*— are driven by ignorance, they will continue to feel desire and hatred. They will inevitably be driven to act on those emotions, creating thereby the karma that perpetuates their continued rebirth, whether that rebirth is "good" or "bad." Virtuous deeds lead to progress only in the sense that they result in temporary happiness within the cycle, never liberation from it.

Ignorance is identified as the conception of self, the sense that there is some autonomous entity who is the owner of mind and body, who is the doer of deeds, who enjoys pleasure and suffers pain, who is the referent of the first person singular pronoun. And this conception is not just a concept. Over the course of billions of lifetimes the sense of self has become so ingrained as to be innate, present not only in humans, it is said, but in all forms of animal life as well, from the flea to the elephant, motivating all behavior, coloring even the sense perception of the world. It sounds something like the selfish gene. But it is immaterial, it is a form of consciousness, driven to survive by self-cherishing, moving forward through time, and through bodies, trying to live forever.

The Buddhist claim is that, as deeply ingrained as the sense of self may seem, it can be uprooted; that by destroying ignorance, desire and hatred become impossible, putting an end to the deeds that drive the cycle of rebirth. Without its fuel, the engine grinds to a halt. And so Buddhist sutras talk about the time when *saṃsāra* is empty, that point in the future, whether hypothetical or not, when there is no more rebirth. And because the world, the physical universe, is also the product of the karma of the beings born there, when there are no more beings to be reborn, the physical universe will cease to exist. No more Nature red in tooth and claw.

According to the theory of evolution, a species avoids extinction through adaptation. If it does not evolve, it will become extinct, and even if its does evolve, it will become extinct eventually. According to Buddhism, the species of the sentient will continue forever, driven by the engine of karma. Yet for Buddhism the highest goal of the species, a goal that can only be conscious, is to seek extinction. Perhaps Buddhism is "life denying" after all, as Christian missionaries have been saying for so long.

As we consider Buddhism and science more generally, we might also compare their epistemologies, their theories of knowledge. In the scientific method, at least in its idealized form, reality or truth (whatever those terms might mean in a given case) has not yet been discovered; hypothesis and experiment are employed to arrive at a truth that is, at that time, unknown, or at least unverified. That truth can change, and has changed many times over the history of science, with a new truth sometimes refining an old truth, sometimes completely replacing it. The image is one of augmentation and revision, moving toward an ever-receding horizon of omniscience, a Theory of Everything.

Over the long and varied development of the Buddhist traditions, there remains the shared belief that the nature of reality was discovered long ago by the Buddha, and before him by the buddhas of the distant past—in the history of our world there have been four or seven or twenty-five previous buddhas, counting Śākyamuni—and that same reality has been understood by all the buddhas in its entirety and its fullness. This reality is represented not as something that the Buddha was the first to discover but as something he rediscovered. In a well-known allegory, the Buddha describes a traveler coming across an ancient city at the end of a dimly visible path through a great forest, a once great city now deserted and in ruins. The traveler informs the king, who restores the city to its former glory. The Buddha is that traveler, discovering the same path to enlightenment that the buddhas of the past had found. In Buddhism, the truth is something that is found, and then lost, and then found again. This is why it is said that the next buddha does not appear in the world until the teachings of the previous buddha have been completely forgotten. The next buddha, who will be called Maitreya, will appear only when there are no more monks, no more monasteries, no more courses on Buddhism or Buddhologists to teach them, no more books about Buddhism. Maitreya will not appear until a Google search for the word "Buddhism" yields the message, "Your search — Buddhism — did not match any documents." In other words, as long as the path to the city of reality remains passable and the city itself remains prosperous, there is no reason for repair. But when the city falls into ruins and the path is overgrown with oblivion, then the path must be cleared again and the city restored. This is what the buddhas do, again and again, over the eons. There is no new knowledge, only the remembrance of what has been forgotten.

If we take the doctrine of karma and the theory of evolution as our test case, perhaps Buddhism is not compatible with science. But if Buddhism could somehow be made compatible with science, what would it look like? The standard cosmology of four island continents surrounding a central mountain, a cosmology incompatible with modern maps, would have to be abandoned. Perhaps it is dispensable. But the demigods live on the slopes of the central mountain, and the gods live on its summit and in the heavens above. The ghosts inhabit a realm five hundred *yojanas* beneath our southern continent. Another five hundred *yojanas* below the realm of the ghosts is the first of the eight hot hells, with the other seven hot hells stacked beneath it. Thus, the six realms of beings would need to be reduced to two: humans and animals. This Buddhism would be a materialist Buddhism, one in which the dichotomy between mind and matter so central to all Buddhist thought and practice would have to be abandoned; mind would become merely an epiphenomenon of matter. And if mind does not pre-cede matter nor persist beyond it, there could be no rebirth. Thus, the Buddha would not have perfected himself over many millions of lifetimes, practicing infinite forms of ten virtues: giving, ethics, renunciation, wisdom, effort, patience, truthfulness, resolution, love, and equanimity. He would not have that bump on his head, multicolored beams of light would not radiate from his mouth when he smiled. His disciples would not rejoice at being reborn at a time when a buddha appeared in the world. The Buddha would not predict the time and the place in the distant future when a lowly disciple would become a buddha. Indeed, the Buddha would be a man, a man who sat beneath a tree and contemplated the law of cause and effect and then spent his life teaching others to live ethically and to understand that there is no self. At the end of his life, he would enter nirvana. Put another way, he would die.

For the scholar of Buddhism, this is obviously a bleaching of the Buddha, a fading of his aura. For the Buddhist, it is a domesti-cation of the dharma, making it *laukika* (something of the world) rather than *lokottara*, something beyond the world.

One might argue that my attempt to understand nirvana in light of scientific views of extinction is simply another attempt to salvage the doctrine of a particular religion because it somehow "sounds like" a scientific category. A Christian could just as easily make the same claim about extinction and paradise. But I drew the parallel not to make yet another claim for the compatibility of Buddhism and science. My claim, or at least my suggestion, is the opposite.

Since the nineteenth century, and continuing with a particular fervor into the twenty-first, there has been a deep desire to make Buddhism into a, perhaps the, post-scientific religion, the one religion that has withstood the critique of science.

But regardless of where science (and, lately, neuroscience) goes, or does not go, we might allow Buddhism to remain a religion, and to be a pre-scientific religion, one in which a king cuts away pieces of his own flesh to feed a hawk and save the life of a dove, until he has no more flesh and he becomes a talking skeleton; where the Buddha utters a single syllable and thousands hear an eloquent discourse in their various native tongues; where someone makes the impossible vow to liberate every being in the universe from suffering. As the great Belgian scholar of Buddhism Msgr. Étienne Lamotte wrote in 1958, "The Buddhist tradition is steeped in the marvelous. Belittled by some schools and exaggerated by others, the marvelous is ubiquitous. We have accepted it as such without attempting to eliminate it in the name of western rationalism. To disregard it would be to offer a caricature of Buddhism and still not attain historical truth. It is not enough to discard legend in order to discern the reality of the facts. By leaving the marvelous the place it has always occupied in the sources, we believe we have given a more faithful image of the mentality of the Buddha's disciples."[15]

It is not the role of the scholar to protect, preserve, and defend the religion that he or she studies. Religions, or at least their adherents, have done that themselves over the centuries. It is the task of the scholar to document and analyze those efforts. Reli-

gions change over time. It is the task of the scholar to document
and analyze that change. To understand what a religion is, it is
essential to understand what it has been at other points in time, to
understand what it is now at other points in space. It is essential to
remember. And it is the task of the scholar to aid in that remembering.

On the night of his enlightenment, the Indian prince, the
prince who would become the Buddha at dawn, sat down to meditate. He meditated all night. It is said that during his first period of
meditation, he had a vision of all of his past lives, a vision referred
to in the texts as "the memory of former abodes." The presence of
memory as the first constituent of the defining experience of the
tradition suggests the importance of memory, of retrospection,
in Buddhism, an importance it has retained throughout its various
developments. There is change, but there is also memory. In Buddhism, it is important to remember.

The Buddhist term for memory is *smṛti* in Sanskrit, *sati* in
Pali, a word that in recent times has come to be rendered in English as "mindfulness." As Buddhism spins off into ever changing
forms, it is essential to be mindful of what it once was. It is also
essential to remember that it has always been spinning.

Thus, rather than seeing Buddhism as compatible with evolution, why not see it as radically incompatible, seeking extinction
rather than survival, seeing persistence only in impermanence,
stressing intention over compulsion, consciousness over matter?
My suggestion is that this incompatibility carries with it a particular power. My suggestion is that we allow the Buddha to remain beyond the world, completely at odds with the world, and
with science.

After Prince Siddhārtha returned from the last of his chariot
rides outside the palace, where he had observed, for the first time

in his twenty-nine years, old age, sickness, and death, he went to his father, King Śuddhodana, and requested permission to go forth from the life of a householder and enter the forest, in ancient India, the place for spiritual pursuits. His father urged him to stay and fulfill the duties of his royal lineage, to postpone his religious quest until he was older. As described by the poet Aśvaghoṣa, the prince replied:

> If you will become a surety for me
> > in four things, O King.
> I will not go to the ascetic grove.
>
> My life shall never be subject to death,
> > disease shall not steal this good health of mine,
> Old age shall never overtake my youth,
> > no mishap shall rob this fortune of mine.

His father replied that this was an extravagant wish, impossible to grant. And then the prince said to his father:

> If that's not possible, don't hold me back;
> > for it is not right to obstruct a man
> Who's trying to escape from a burning house.[16]

It was escape from the burning house of birth, death, and rebirth that the Buddha sought, and according to the tradition, he found it, and then taught it to others, setting forth a path that leads to the cessation of physical form and to the cessation of consciousness, a permanent state of freedom which is ultimately an absence of life, at least life as it has been understood by science. Thus, far from teaching a dharma compatible with Darwin's theory of natural selection, it is perhaps more accurate to regard the Buddha as a counter-evolutionary, actively seeking the extinction of the human race, and indeed of all species, through the eradication of the selfish gene.

A Primer on Buddhist Meditation

An essential element in the modernization of meditation is the claim that a particular form of practice is not "religious." It may be "Buddhist," in the sense that its origin is attributed to the Buddha himself, but it is not part of a religious practice, not bound by time or by culture. Yet, although meditation may seek the timeless, its practice originated in time. As meditation, and especially what is described as Buddhist meditation, becomes an increasing focus of scientific research, it is perhaps useful to pause for a few moments to consider what meditation has meant in the history of Buddhism.

It is not the case, as it is sometimes claimed, that meditation is to Buddhism what prayer is to Christianity. Over the course of its long history across Asia, Buddhist laypeople have not been expected to practice meditation, nor have they done so. Meditation has traditionally been a practice confined to monks and nuns, and even then, not all monks and nuns have meditated, especially on a daily basis. Early Buddhist instructions for the design of a monastery state that a quiet area should be set aside for the meditating monks, suggesting that meditation was but one of a number of specializations in the traditional Indian monastery. There has also been long-standing debate over the importance of meditation versus the importance of the study of the scriptures. Around 35 B.C.E., a controversy developed in Sri Lanka as to whether the practice of meditation or knowledge of doctrine was more crucial to the preservation of Buddhism. The proponents of the latter position prevailed.

Nevertheless, since the early days of the tradition, meditation has been regarded as the practice par excellence in Buddhism, and the many stages along the Buddhist path, with attainments both mundane and transcendent, are to be achieved through the practice of meditation. The radical critiques of formal meditation that have occurred at various moments in the tradition (one thinks immediately of certain forms of Zen) derive their radical quality from the fact that they challenge a practice that is so central to the tradition, in ways both symbolic and real.

The most common depiction of the Buddha, across the Buddhist world, is in the meditative posture, and the many accounts of the defining moment for the tradition, the Buddha's enlightenment, say that he achieved enlightenment at dawn after a night of solitary meditation, seated beneath a tree. Exactly what his meditation practice was that night has been the subject of considerable commentary and conflicting claims over the long history of Buddhism. In the centuries after his death, an extensive literature on the theory and practice — and between these two, especially on the theory — of meditation developed in India.

As is so often the case in the study of Buddhism, one of the first issues to consider is that of translation. What Sanskrit term (or terms) is translated into English by "meditation," and what is lost in translation? First, it is important to note that there is no single Buddhist term that can be accurately translated as "meditation," at least as that term is popularly understood in the West. The term most commonly rendered as "meditation" is *bhāvanā*, derived from the Sanskrit root *bhu*, "to be" or "to become," and having a wide range of meanings including cultivating, producing, manifesting, imagining, suffusing, and reflecting. It is the first sense, that of cultivation, which is often mentioned by commentators, noting the importance of the sustained development of particular states of mind. And yet, just as the term "cultivation" has a wide range of meanings in English, the term *bhāvanā* in Buddhism encompasses much more than the silent practice of mental absorption; it can include studying doctrine, memorizing sutras, and chanting verses to ward off evil spirits.

Another Sanskrit term that is translated as "meditation" is *dhyāna*, although it is also rendered as "concentration" or "absorption." Unlike *bhāvanā*, the more generic term, *dhyāna* has a specific meaning in Buddhism, to which we will turn shortly. Finally, there is *samādhi*, which has a sense of joining together or bringing into harmony. In the context of Buddhist practice, it is a general term for various states achieved through specific practices, some of which might be termed contemplative.

Over the course of the centuries, thousands of meditative practices have developed across the Buddhist cultures of Asia, and a great many of these deviate widely from the notion of a monk seated cross-legged absorbed in a state of silent insight into the ultimate truth. Some forms of Buddhist meditation seek to produce visions of distant buddhas and their marvelous lands, allowing the meditator to see these buddhas face to face and hear their teachings. Others involve the recitation of mantras to summon a deity in order to request boons and blessings. In China, much meditation was devoted to the practice of repentance. One might protest that such practices are not meditation but rather forms of ritual. However, it might be argued that all forms of meditation over the long history of Buddhism are ritual practices.

When we turn from what was done in meditation in Buddhist monasteries to what was said about meditation in Buddhist texts, we arrive at a traditional division that provides a useful framework within which to place many, although not all, forms of Buddhist meditation. It is a division into those meditations whose aim is serenity and those whose aim is insight.

In its classical formulation, the practice of Buddhist meditation begins with a number of premises. The first is that the ordinary human mind is pervaded with all manner of pollutants, described with terms like affliction, hindrance, and contaminant, factors that both obscure the vision of things as they are, and motivate various negative deeds that cause suffering for oneself and for others, both now and in the future. It is believed, however, that these contaminants can be cleansed, and the practice of meditation is the chief means for doing so. That is, the contaminated mind can purify itself of contamination. Whether this can be fully achieved without the ultimate cessation of the mind itself is a question that would become central to the development of Buddhist thought. A further premise, however, is that the ordinary human mind is incapable of successfully undertaking the process of purification because it is in a state of almost perpetual distraction, racing uncontrollably from one object to the next, unable to focus on anything for a sustained period of time.

The primary aim of serenity meditation is to bring the mind under control. This is done by focusing the mind on a particular object and forcibly returning it to that object whenever it strays. The goal is to develop the ability to place the mind one-pointedly on an object and have it remain there, effortlessly, for as long as one chooses. Many things can serve as the object of concentration; there is a traditional list of forty, including such things as the breath, the foulness of the human body, compassion, and the qualities of the Buddha. The power of concentration can be increased to higher and higher states, over nine levels, each more sublime than the next. These states, and their attainment, are not considered unique to Buddhism; they are said to be accessible through the practice of other religions as well.

One of the persistent issues in the discourse of Buddhism and science is how much of traditional Buddhist doctrine should be relegated to the category of myth. One might imagine that this problem does not extend into the practice of meditation, yet it does. Perhaps the most immediate manifestation of this problem is the fact that the nine levels of concentration are also nine places of rebirth; indeed they provide the basic structure of the Buddhist universe. Here, the beings of the universe are categorized by their powers of concentration. These nine levels are divided into three realms, called the Realm of Desire, the Realm of Form, and the Formless Realm. Humans and animals (as well as hell beings, ghosts, and some types of gods) inhabit the Realm of Desire, so called because its inhabitants constantly crave the pleasures of sense experience—of pleasing forms, sounds, fragrances, flavors, and things to touch—and their concentration is weak, so weak, it is said, that it is impossible to achieve enlightenment with such a mind. This does not mean that humans are doomed to remain in *saṃsāra*, but that they must develop an extraordinary level of concentration in order to escape from it. Specifically, in order to achieve enlightenment and liberation, it is necessary to develop a level of concentration beyond that of the Realm of Desire. Minimally, one must develop something called "access concentration."

But it is also possible, and repeatedly recommended in the scriptures, to develop higher levels of concentration of the Realm of Form and of the Formless Realm.

The Realm of Form and the Formless Realm each have four levels; in the Realm of Form, they are simply called the first, second, third, and fourth concentrations (in ascending order). In the Formless Realm, the levels are named for the object of contemplation of the beings who inhabit them: Infinite Space, Infinite Consciousness, Nothingness, Neither Existence nor Non-existence. In an interesting doctrinal twist, it is said that a meditator who achieves a specific level of concentration of the Realm of Form or the Formless Realm as a human will be reborn in that level in their next lifetime; that is, if one reaches the level of concentration of the First Concentration in one's meditation practice, for example, one will be reborn there after death. In Buddhist cosmology, the eight levels of the Realm of Form and the Formless Realm are heavens and the beings reborn there are gods, although in the Formless Realm, as the name suggests, they have no bodies. They are only minds, contemplating Infinite Space, for example, for millennia.

Each of the levels of concentration of the Realm of Form and the Formless Realm is a profound state of peace, and the Buddha often exhorts monks to master them. He himself did so, and demonstrated that when he died. As he lay on his right side between two trees, his mind moved from the Realm of Desire up through the four levels of the Realm of Form and the four levels of the Formless Realm, then back down to the Realm of Desire, then back up to the Fourth Concentration of the Realm of Form, from which he entered nirvana. However, and this is a very important point, the development of concentration does not in itself constitute liberation from rebirth, nor is rebirth in the Formless Realm a state beyond the cycle of birth and death. Although the lifetime is long and free from the ordinary sufferings of the world, the consciousness that inhabits that realm must eventually be reborn elsewhere. Indeed, in the context of the path to liberation, the four levels of the Formless Realm are sometimes portrayed as diversions, and the saints of other religions are said to mistake them for the ultimate state of liberation. The four levels of the Realm of Form, although lower in strength of concentration, figure more prominently in prescriptions for the path.

Concentration in and of itself is incapable of bestowing liberation, or, stated more technically, concentration can suppress the operation of the various contaminants that pervade the mind, but it cannot permanently remove them. For that, wisdom is required, and hence the other major category of meditation consists of those practices that cultivate insight. This insight is specific. It is insight into the nature of reality, described in early texts as the understanding that everything bears three marks: impermanence, suffering, and no self. More generally, it is a single insight: that there is no self, that there is nothing permanent, enduring, independent, indivisible, autonomous among the physical and mental elements that constitute the person and that might be called "the self." The precise nature of this particular absence—this absence of self whose understanding constitutes the highest wisdom, and the liberating insight, in Buddhism—is the subject of endless commentary across the tradition. Yet, there is the consistent claim that the understanding of this absence is wisdom, that this is insight.

The first category of meditation, serenity meditation, tends to encompass what people often imagine meditation to be: single-pointed focus on a single object, a form of trance in which conceptual thought is suspended, resulting in a state of inner calm. The second category, analytical meditation, is different. Here, meditation often entails a process of reflection, and even discursive reasoning, in which a thoroughgoing mental search is launched by the mind in an attempt to find among the elements of body and mind something that lasts longer than an instant without passing away, something that might be the self. The Buddhist claim is that no such self is to be found.

The intellectual understanding that there is no self, however, is not sufficient to bring about the destruction of the defilements. That insight must be empowered by a level of strength beyond that of ordinary conclusions. In order for the defilements to be destroyed, that insight must be endowed with a level of concentration beyond that of the Realm of Desire; it must be that of the Realm of Form, or at least that of "access concentration." Analysis must be stabilized. Serenity must be conjoined with insight. This is how concentration and insight work together to destroy the causes of suffering. It is sometimes said that insight is the sharp ax and concentration is the strong arm. Together they can cut down the tree of suffering at its roots; individually the ax and the arm cannot fell it.

Buddhist scholastics in India described in precise detail how the mind endowed with serenity and insight uproots suffering and its causes. It is a process that, at least at first sight, seems hopelessly complicated, and quite foreign to what, for want of a better term, might be called religious experience. The process of purification, according to one influential description, takes place over the course of five stages, called the five paths, stages that may occur over the course of several lifetimes. On the first path, called the path of accumulation, the practitioner develops morality, merit, and meditative skill, including preliminary understanding of no-self. The means for understanding that there is no self is sometimes identified as the "four establishments of mindfulness," to which we will return below. The second path, the path of preparation, begins when the meditator gains a conceptual (rather than direct) insight into the reality of no self, at a level of concentration beyond that of the Realm of Desire. The insight at this level, however, is insufficiently strong to destroy any of the defilements. The meditator contemplates what are known as "the sixteen aspects of the four truths," that is, four qualities of each of the four noble truths of suffering, origin, cessation, and path. For example, the four aspects of the first truth, suffering, are impermanence, misery, emptiness, and no self; the four aspects of the fourth truth, the path, are path, suitability, progress, and achievement. It is on the third of the five paths, called the path of vision, that the meditator begins to destroy, or in the language of the tradition, abandon, the defilements. This occurs over sixteen stages of realization, four for each of the four truths. These sixteen stages of realization, however, are different from the sixteen aspects of the four truths. In the first moment, called "forbearance," the defilements of the Realm of Desire associated with the truth of suffering are abandoned. This is followed by a second moment, called "knowledge," which is the understanding that the defilements of that level with regard to the truth of suffering have been abandoned. This is followed by the third moment, called "subsequent forbearance," in which the defilements of the Realm of Form and the Formless Realm associated with the truth of suffering are abandoned. The fourth moment, called "subsequent knowledge," is the understanding that the defilements of those two realms (again, with regard to the truth of suffering) have been abandoned. This fourfold sequence is repeated for the three other truths—origin, cessation, and path—resulting in sixteen stages.

On the path of vision, however, only the more superficial defilements are destroyed, those associated particularly with the mistaken belief in self. As a result, the meditator destroys all causes for future rebirth as an animal, ghost, or hell being, but is not liberated from rebirth altogether and may still be reborn as a human or god. The more deeply rooted defilements are destroyed over the course of the next path, called the path of meditation.

Here, the numbers begin to pile up. It is said that for each of the nine levels of the three realms of rebirth—the Realm of Desire (with one level), the Realm of Form (with four levels), and the Formless Realm (with four levels)—there are nine gradations of defilements, from the most coarse to the most insidious. Nine times nine makes eighty-one levels of defilement to be destroyed. As was the case with the path of seeing, these must be destroyed in a two-step process: the actual destruction of the particular defilement and the knowledge that it has been destroyed, a process likened to throwing a thief out of your house and then locking the door behind him. Eighty-one times two is one hundred and sixty-two "moments" of the abandoning of defilements. This process, which takes place over the course of the path of meditation, may again occur over several lifetimes. However, when the one hundred and sixty-second stage is reached, and the most subtle of the subtle defilements associated with the ninth level—that is, the fourth absorption of the Formless Realm—has been abandoned, one is liberated from rebirth.

This view of the Buddhist path is obviously sharply at odds with the common view of "meditation," Buddhist or otherwise, immediately raising rude questions about how one would know that one had just abandoned defilement #44 (which would be the eighth of the nine levels of defilement of the fourth concentration of the Realm of Form). Or, more generally, how such a rubric could have been derived from anyone's experience in meditation. However, it is important to note that this view of the Buddhist path remained orthodox in India over the course of a millennium. This is not to say that there were not modifications. In some Mahayana schools, for example, the object of wisdom became simply emptiness, rather than the sixteen aspects of the four truths; the eighty-one levels of defilement remained.

The Mahayana also added its own baroque embellishments to the temple of Buddhist meditation. One of these took the form of one of the terms that is sometimes translated into English as "meditation," *samādhi*. These are meditative states, perhaps visionary states, that the bodhisattva enters, from which all manner of magical powers become possible. A ninth-century Tibetan lexicon of Buddhist terminology lists one hundred and eighteen different forms of *samādhi*, drawn from one genre of Mahayana literature alone, the Perfection of Wisdom sutras. These include states with names like "sending out rays of light," "seeing in all directions," "not forgetting," "pervading space," "causing joy," "lamp of lightning," "the equality of all things," "inexhaustible casket," and "completely removing the faults of the body." Among the most famous is one called "the *samādhi* of the direct encounter with the buddhas of the present." According to Mahayana doctrine, although the Buddha of our world is no longer in the world, at present there are other buddhas in other worlds of the universe. Those who develop this *samādhi* have the ability to meet with these buddhas face to face and receive their teachings. Scholars speculate that such visions yielded new scriptures that came to be regarded by their adherents as canonical, long after the death of the historical Buddha.

We thus see that the meanings of meditation are many in Buddhism, and their importance is pervasive. Even for those traditions in which the practice of seated meditation described above is not emphasized, the significance of such practice looms large.

Thus, in the Pure Land traditions of Japan, seated meditation is eschewed, not because it is unimportant but because in the degenerate age in which we live, beings lack the necessary requisites to practice it effectively. In China, Dahui (1089–1163) was a harsh critic of what was called "silent illumination Chan," that is, prolonged sessions of silent seated meditation in which one sought to cease mental activity in order to reveal the original purity of the mind. Arguing that such a practice resulted only in calluses on the buttocks, he instead advocated the active wrestling with koans, statements such as, "Q: What is the Buddha? A: Three pounds of flax." "Q: What is the Buddha? A: A dried piece of shit." The tantric practices that developed in India, condemned by Victorian scholars for their sexual symbolism, in fact contained highly sophisticated visualization practices that require both serenity and insight, including a practice in which the meditator gains an understanding of emptiness and then causes that understanding, that mind, to appear in vivid detail as a *maṇḍala*, an opulent palace with a buddha seated on the central throne. One of the signs of success in such practice is the ability to shrink such a visualized palace down to the size of a mustard seed and remain in one-pointed concentration on it for extended periods of time, with each of the details of the palace (such as the strings of jewels that adorn the pillars) and of the deities who inhabit it (such as the color of their eyes) vividly appearing.

This very brief survey of the meaning of meditation in Buddhism is provided to suggest that it is inaccurate to assume that Buddhist meditation is encompassed by something called mindfulness, which has come to represent Buddhist meditation in many conversations, including those in the domain of science, in recent years.

The Sanskrit term that is so often rendered as "mindfulness" in English is *smṛti*, whose most basic meaning is "memory." In Hinduism, one of the categories of sacred literature is called *smṛti*, that which is to be remembered. The term also occurs frequently in Buddhism. In the Buddhist description of the ordinary functioning of the mind, *smṛti* is that factor which prevents the mind from forgetting something. In the context of meditation practice, there are many cases in which the term is best rendered simply as "memory" or "recollection." Thus, as mentioned in the previous chapter, on the night of his enlightenment, during the first watch of the night, it is said that the Buddha remembered all of his past lives. Among the traditional forty objects that can serve as the focus for developing concentration, one finds the memory or recollection of death, that is, to remember that death is certain but the time of death is uncertain. Another meditation is the memory or recollection of the Buddha, in which one recalls the Buddha's epithets in the following famous phrase, "The Blessed One is such since he is accomplished, fully enlightened, endowed with clear vision and virtuous conduct, sublime, the knower of worlds, the incomparable leader of humans to be tamed, the teacher of gods and humans, enlightened and blessed." In each of these cases, this practice is called *smṛti*.

There are other contexts, however, in which "memory" is not the best translation of *smṛti*. In that same list of forty objects for developing concentration, one finds the breath. Since we do not need to remember to breathe, it would seem that the practice would instead be to focus the mind on the breath, to be mindful of breathing. And so the Sanskrit term has come to be most commonly rendered in English as "mindfulness." One of the first British scholars to render *smṛti* (or *sati*) into English was the Wesleyan missionary to Ceylon (and critic of Buddhism) Daniel J. Gogerly (1792–1862), who in a paper read to the Ceylon Branch of the

Royal Asiatic Society in 1845 translated it simply as "meditation."[1] A few years later, his colleague Robert Spence Hardy (1803–1868) translated the term as "conscience," calling it "the faculty that reasons on moral subjects; that which prevents a man from doing wrong, and prompts him to do that which is right."[2] According to the Oxford English Dictionary, although the term "mindfulness" in the sense of attention or purpose appears as early as 1530, the first occurrence of the term in the context of Buddhism is in 1889 by Sir Monier Monier-Williams (1819–1899), who was born in India as the son of a British colonial officer and went on to become Boden Professor of Sanskrit at Oxford. It appeared in his book *Buddhism, in Its Connexion with Brahmanism and Hinduism, and in Its Contrast with Christianity*. In fact, however, its initial occurrence appears to have been in 1881 in a book called *Buddhist Suttas*, part of Max Müller's landmark Sacred Books of the East series. The translator was Thomas W. Rhys Davids (1843–1922), a former British colonial officer in Sri Lanka who went on to become the most celebrated Victorian scholar of Buddhism. In his list of the eightfold path, he called the seventh element, "Right mindfulness; the watchful, active mind."[3] In 1890, Rhys Davids again used the term in his translation of *The Questions of King Milinda*, also published in Max Müller's series. In this renowned text, a Buddhist monk answers the questions of a Greek king. In the translation by Rhys Davids, the monk explains to the king what mindfulness is: "As mindfulness springs up in his heart, O king, he searches out the categories of good qualities and their opposites, saying to himself: 'Such and such qualities are good, and such bad; such and such qualities helpful, and such the reverse.' Thus does the recluse make what is evil in himself to disappear, and keeps up what is good. That is how keeping up is the mark of mindfulness."[4] It is clear from this passage that mindfulness is "judgmental," contrary to what more modern proponents would claim.

In Buddhist meditation, *smṛti* (which to avoid further confusion will be rendered as "mindfulness" from this point) is not so much a form of meditation as a factor necessary for success in any form of meditation, "like a seasoning of salt in all sauces, like a prime minister in all the king's business." In a list of thirty-seven factors conducive to enlightenment, it occurs fives times, including as the seventh element of the eightfold path. Among the more important lists in Buddhism is the three trainings: in ethics, in meditation (here, *samādhi*), and in wisdom, the three things necessary for enlightenment. Ethics is the restraint of body and speech, meditation is the restraint of mind through the development of concentration, wisdom is insight into the nature of reality. Mindfulness is included under the second of these three trainings, the training in meditation.

It is mindfulness that places the mind on the chosen object of meditation and returns the mind to the object when it wanders. Mindfulness prevents distraction. As a well-known meditation instruction says, "Tie the wild elephant of the mind to the post of the object with the rope of mindfulness." Mindfulness is also said to protect the mind from the intrusion of unwanted elements — whether they be from the senses or from thoughts — like a guard at the door.

The term "mindfulness" figures prominently in a famous discourse of the Buddha, entitled the *Satipaṭṭhana Sutta*, the *Discourse on the Foundations of Mindfulness*. Here, the Buddha sets forth what he calls the *ekāyana magga*, "the only path" or "the one way." Four objects of mindfulness are prescribed. The first is the mindfulness of the body. The second is the mindfulness of feelings, which here refers to physical and mental experiences of pleasure, pain, and neutrality. The third is the mindfulness of the mind, in which one observes the mind when influenced by different positive and negative emotions. The fourth is called the mindfulness of dharmas, which here means the contemplation of several key categories, including the constituents of mind and body and the four noble truths.

The first of the four, the mindfulness of the body, involves fourteen exercises, beginning with the mindfulness of the inhalation and exhalation of the breath. This is followed by mindfulness of the four physical postures of walking, standing, sitting, and lying down. This is then extended to a full awareness of all activities. Thus, mindfulness is something that is meant to accompany all activities in the course of the day, and is not restricted to formal sessions of seated meditation. This is followed by mindfulness of the various components of the body, a rather unsavory list that includes fingernails, bile, spittle, and urine. Next is the mindfulness of the body as composed of the four elements of earth (the solid), water (the liquid), fire (the warm), and air (the empty). This is followed by what are known as the "charnel ground contemplations," mindfulness of the body in nine successive stages of decomposition of a human corpse.

This practice of the mindfulness of the body is intended to result in the understanding that the body is a collection of impure elements that arise and cease in rapid succession, utterly lacking any kind of permanent self. That is, the body, and indeed all conditioned things, are marked by three qualities: impermanence, suffering, and no self.

The second of the four establishments of mindfulness, the mindfulness of feeling, observes the various feelings of pleasure, pain, and neutrality that occur in the mind and body, noting when and where they arise and when and where they vanish. The third of the four establishments of mindfulness gives similar attention to the various types of mental states and emotions, such as lust, hate, delusion, and distraction.

The fourth and final establishment of mindfulness takes dharmas as its object. Here the term *dharma* means "phenomena," or the constituents of the world. This section takes up some of the standard lists of Buddhist philosophy: the five hindrances, the five aggregates, the six bases, the seven enlightenment factors, and the four truths. In each case, the meditator recognizes each for what it is: its presence, its absence, how it arises, how it disappears. In the case of virtuous states of mind, the meditator comes to understand how to cultivate them and how to sustain them once they have been cultivated.

At the end of this discourse, the Buddha makes a powerful claim for the efficacy of the practice, calling it "the direct path for the purification of beings," which leads to nirvana in as little as seven days. It is a version of this meditation that today is prescribed (according to a poster at a local hospital) to "manage stress; quiet habitual thoughts; boost disease resistance; maintain overall health; manage high blood pressure, sleep disorders, lifestyle changes, physical and emotional pain." It is also prescribed for depression.

As noted above, over the long history of Buddhism across Asia, laypeople have rarely practiced meditation. This began to change, at least in Burma, about a century ago. When Burma came under the control of the British in 1885, the king was deposed, and the *saṅgharāja*, the "king of the community," the monk appointed by the king to oversee the Buddhist monastic community, lost his authority. This led to a state of disorganization, and a number of individual monks took on the task of preserving the dharma. Believing that the presence of the British was a harbinger of the degenerate age, a monk named Ledi Sayadaw (1846–1923) sought to ensure the preservation of the Buddhist teaching by spreading it as widely as possible. He first began teaching Buddhist philosophy, in the past a specialty of scholar monks, to laypeople. He later began to teach meditation, in the past a specialty of forest monks, to laypeople as well. Another monk, U Nārada (1870–1955), chose as his central text the Buddha's *Discourse on the Foundations of Mindfulness* summarized above. It is unclear how widespread this particular practice of mindfulness was among Burmese monks in the period prior to the British and how much was U Nārada's innovation. It is clear that he made the instructions simpler than those provided in the rather abstruse discourse of the Buddha by, among other things, focusing the practice especially on the first of the four foundations of mindfulness, the mindfulness of the body, beginning with the mindfulness of the breath. He taught this practice to the laity and encouraged the establishment of meditation centers for group practice; meditation had generally been a solitary monastic practice in Burma up to this point.

The practice of meditation, insight (*vipassanā*) meditation, based on the *Discourse on the Foundations of Mindfulness*, would become something of a national, and nationalistic, craze in Burma, with hundreds of meditation centers and thousands of meditators, as Burma achieved its independence, first from the Japanese with the end of the Second World War and then from the British, in 1948. In 1954, a noted monk from Sri Lanka, Nyanaponika Thera (1901–1994, born Siegmund Feniger in Germany), traveled to Burma for a meeting, where he received instruction on what has come to be known as "the Burmese method." He went on to write a book about it, *The Heart of Buddhist Meditation*, bringing the practice to the English-speaking world. His enthusiasm was so great that he represented it as a kind of universal practice, one that did not require that one ascribe to Buddhism, or to any religion: "It is a vital message for all: not only for the confirmed follower of the Buddha and his Doctrine (Dhamma), but for all who endeavor to master the mind that is so hard to control, and who earnestly wish to develop its latent faculties of greater strength and greater happiness . . . for that vast, and still growing, section of humanity that is no longer susceptible to religious and pseudo-religious sedatives, and yet feel, in their lives and minds, the urgency of fundamental problems of a non-material kind calling for solution that neither science nor religions of faith can give."[5]

Since the publication of this book in 1965, there has been a steady increase in the occurrence of the word "mindfulness" in English-language publications, a word that rarely appeared prior to 1950, with an unbroken ascent since the early 1980s.

From Burma, mindfulness meditation spread to other countries in Southeast Asia and to Sri Lanka and then to India, where youthful seekers from Europe and America enrolled in meditation retreats. From India, it came to America. The "mindfulness" that is now taught in hospitals and studied in neurology laboratories is thus a direct result of the British overthrow of the Burmese king.

CHAPTER FOUR

The Death of the Scientific Buddha

Chapter 79, entitled "The Prairie," is one of the cetological chapters of *Moby-Dick*. Here, Melville, in the voice of "unlettered Ishmael," describes the head of the Sperm Whale:

> To scan the lines of his face, or feel the bumps on the head of this Leviathan; this is a thing which no Physiognomist or Phrenologist has as yet undertaken. Such an enterprise would seem almost as hopeful as for Lavater to have scrutinized the wrinkles on the Rock of Gibraltar, or for Gall to have mounted a ladder and manipulated the Dome of the Pantheon. Still, in that famous work of his, Lavater not only treats of the various faces of men, but also attentively studies the faces of horses, birds, serpents, and fish; and dwells in detail upon the modifications of expression discernible therein. Nor have Gall and his disciple Spurzheim failed to throw out some hints touching the phrenological characteristics of other beings than man. Therefore, though I am but ill qualified for a pioneer, in the application of these two semi-sciences to the whale, I will do my endeavor. I try all things; I achieve what I can.

The geographical and archaeological references remain familiar, but the science does not. For Ishmael, the term "physiognomy" does not refer simply to the general appearance of an object or organism, as the term is used today, but rather to its more literal meaning of "judging nature," specifically through the study of a person's face. This ancient science, said to have been championed by Aristotle and Pythagoras, was revived by the Swiss poet and pastor Johann Kaspar Lavater (1741–1801) in his treatise *Physiognomical Fragments for the Promotion of the Knowledge and Love of Man* (Physiognomische Fragmente zur Beförderung der Menschenkenntnis und Menschenliebe), published in four volumes between 1775 and 1778. It enjoyed great influence in Europe.

Gall is Franz Joseph Gall (1758–1828), the German physician who developed what came to be known as the science of phrenology; Spurzheim was his best known follower, the German physician Johann Gaspar Spurzheim (1776–1832). In 1809, Dr. Gall published (in German) *The Anatomy and Physiology of the Nervous System in General, and of the Brain in Particular, with Observations upon the Possibility of Ascertaining the Several Intellectual and Moral Dispositions of Man and Animal, by the Configuration of their Heads.* There, he argued that the brain was the seat of the mind, and that various elements of human character, both good and evil, were localized in specific areas of the brain. He enumerated twenty-seven such faculties, including the impulse to reproduce, love for one's children, self-defense, murder, cunning, memory of words, metaphysical capacity, poetic talent, and religious sentiment. He argued that the shape of the brain was determined by the relative size, and hence strength, of these various functions. And he argued—and this is what he is most famous for—that the shape of the brain determines the shape of the skull, such that the various elevations and depressions on the surface of the cranium can be read by a skilled physician to determine a person's character and intelligence. Phrenology became very popular in Europe, especially Britain, in the first half of the nineteenth century, despite scathing criticism as early as 1815 in the *Edinburgh Review*. Today, it is considered a classic example of a pseudoscience; indeed, Karl Popper identified phrenology (along with astrology) as a pseudo-science when setting forth his notion that falsifiability is what de-marcates science from pseudoscience.

Promoted by the American Fowler brothers, Lorenzo and Orson, phrenology remained popular in Britain and America throughout the nineteenth century. By the middle of the century, however, it had been widely disparaged by scholars.

On August 26, 1873, twenty-one years after the publication of *Moby-Dick*, a great debate took place between Christianity and Buddhism, between a Methodist minister and a Buddhist monk, in the town of Panadure in Sri Lanka. Each of the parties in the debate would claim that his religion was compatible with science, although what they meant by science seemed to differ. For the Buddhist, "science" often seemed to mean astrology. He noted that the great astrologers, who were able to predict the date of a person's death, always invoked the Buddha. For the Christian, "science" meant modern astronomy, geography, and technology, that is, what was considered the best science of the day.

Samuel Langdon (1847–1908), a Wesleyan minister from Britain who reported on the debate in the *Quarterly Letters Addressed to the General Secretaries of the Wesleyan Methodist Missionary Society*, looked out on the sea of shaven heads of the Buddhist monks in the audience and wrote, "The sight of so many heads, some very fine ones, so closely shaven, would have been a rare treat for us if we had been disciples of Gall and Spurzheim."[1]

In fairness to Samuel Langdon, it is possible that his comment about the sea of Buddhist craniums at Panadure was made in jest. But there is also something cautionary about his remark. Over the course of the history of science, what was once science has become pseudoscience. And some of those pseudosciences have been linked to Buddhism. Henry Steel Olcott's *Buddhist Catechism*, first published in 1881, contains a chapter titled "Buddhism and Science." There we find a number of questions and answers, one of which has to do with the Buddha's ability to create illusions. Colonel Olcott writes, "360. *Q. Is this branch of science well known in our day?* A. Very well known; it is familiar to all students of mesmerism and hypnotism."[2]

As I noted in Chapter One, in the long history of the discourse of Buddhism and science, what has been meant by Buddhism has changed over the decades. In the beginning, Buddhism was the original Buddhism postulated by European Orientalists, a Buddhism that then came to be identified with the Theravada traditions of Sri Lanka and Southeast Asia, or at least with their Pali canon. In the period after the Second World War, Buddhism became Zen, especially as it was represented by D. T. Suzuki. During the 1960s and '70s, Buddhism was often the Madhyamaka philosophy of Nāgārjuna, and the doctrine of emptiness. Over the past two decades, the Buddhism in dialogue with science has largely been Tibetan Buddhism, a form of Buddhism that just a century ago was regarded as a form of superstition so degenerate that it did not deserve the name Buddhism, but was referred to instead as Lamaism. A century later, the figure once known to Europeans as the Grand Lama of Lhasa, shrouded in mystery for so long, holds annual seminars with some of the leading scientists in the world.

The referent of "science" has also changed. Although quantum physics and cosmology still capture attention in some quarters today, the greatest energy is being directed toward neuroscience, and especially research on meditation.

There are a host of questions to be considered surrounding

this newest domain of Buddhism and science. Rather than pointing to affinities between particular Buddhist doctrines and particular scientific theories, research on meditation has sought to calculate the physiological and neurological effects of Buddhist meditation. Such research would seem to introduce a welcome empirical element to the Buddhism and science discourse.

The assertions being made in this domain are qualitatively different from the assertion that the Buddha understood the theory of relativity. The claim here is that Buddhist meditation works. However, in order to understand the laboratory findings, such a claim requires that one first identify what is "Buddhist" about this meditation, describe what the term "meditation" encompasses in this case, and perhaps the most difficult task: explain what "works" means, especially in the context of the exalted goals that have traditionally been ascribed to Buddhist practice.

In 2007, the Agency for Healthcare Research and Quality, a branch of the U.S. Department of Health and Human Services, published a report on meditation at the request of the National Center for Complementary and Alternative Medicine. The purpose of the report was to review and synthesize the state of research on a variety of meditation practices by examining 813 previous studies of meditation. It is a long report, 472 pages. The report concludes:

> The field of research on meditation practices and their therapeutic applications is beset with uncertainty. The therapeutic effects of meditation practices cannot be established based on the current literature. Further research needs to be directed toward the ways in which meditation may be defined, with specific attention paid to the kinds of definitions that are created. A clear conceptual definition of meditation is required and operational definitions should be developed. The lack of high-quality evidence highlights the need for greater care in choosing and describing the interventions, controls, populations, and

outcomes under study so that research results may be compared and
the effects of meditation practices estimated with greater reliability
and validity. Firm conclusions on the effects of meditation practices
in healthcare cannot be drawn based on the available evidence. It is
imperative that future studies on meditation practices be rigorous in
design, execution, analysis, and reporting of the results.[3]

Such a conclusion is surely cautionary. However, perhaps
more interesting are the kinds of studies one finds surveyed in the
report. Meditation has been tested for its benefits for weight loss,
for lowering blood pressure, for lowering cholesterol, and for re-
ducing substance abuse. That is, meditation is regarded in these
studies as a therapy for stress reduction. Indeed, one of the forms
of meditation examined in the federal study is MBSR, Mindfulness Based Stress Reduction, which seeks to induce a form of
awareness that focuses on the present moment, observing "the
unfolding of experience, moment to moment."

But is stress reduction a traditional goal of Buddhist meditation? Perhaps one might think of nirvana as the ultimate state of
stress reduction, the complete cessation of all physical and mental
processes. We recall that in *Beyond the Pleasure Principle*, Freud
used the term "Nirvana principle" to refer to the tendency of the
psyche to reduce the quantity of internal and external excitation as
much as possible, even to zero. Elsewhere, he linked the Nirvana
principle to the death instinct.

A glimpse at any number of forms of Buddhist meditation, however, suggests that stress reduction is often not the aim.
We might consider one of the most common teachings of the
Nyingma or "Ancient" sect of Tibetan Buddhism, called the four
ways of turning the mind away from *saṃsāra* (*blo ldog rnam bzhi*).
These are part of the so-called preliminary practices (*sngon 'gro*),
meditations that must be completed in order to receive tantric ini-

tiation. Versions of these practices are found among all four of the major sects of Tibetan Buddhism.

The first of these is meditation on the rarity of human birth, how, among the beings that populate the six realms of rebirth, those reborn as humans with access to the Buddha's teaching are incredibly rare. Among the many metaphors for the difficulty of attaining such a human birth, one of the more poetic is that of a single blind tortoise swimming in a vast ocean, surfacing for air only once every century. On the surface of the ocean floats a single golden yoke. It is rarer, said the Buddha, to be reborn as a human with the opportunity to practice the dharma than it is for the tortoise to surface for its centennial breath with its head through the opening in the golden yoke. The second meditation is on the certainty of death and the uncertainty of the time of death, the recognition that one will definitely die, yet the time of death is utterly indefinite. If this is the case, if one cannot say with certainty which will come first—the next moment or the next lifetime—what pleasure is to be found in the world? This is the subject of much Buddhist poetry.

> Do those who fall to earth from a mountain peak
> Find happiness in space as they rush to destruction?
> If we constantly race toward death from the time of birth,
> How can beings find happiness in the time in between?[4]

The third preliminary practice is to meditate on the workings of the law of karma, how negative deeds done in the past will always ripen as suffering and how over the beginningless cycle of rebirth each of us has committed countless crimes. The prospect of eternal suffering lies ahead. And what are those sufferings? The fourth meditation is on the faults of *saṃsāra*, visualizing in detail the tortures of the eight hot hells and the eight cold hells, the four neighboring hells, and the various trifling hells; the horrible hunger and thirst suffered by ghosts; the sufferings of animals, the sufferings of humans that we know so well, even the sufferings of gods. For in Buddhism, the gods also suffer. It is said that certain signs attend the death of a god in heaven. His grand throne becomes uncomfortable, he begins to perspire, the garlands of flowers around his neck begin to wilt, his servants seem reluctant to approach him, his palace becomes dusty. At that moment, it is said, the god has a vision of his next lifetime and, because a god will inevitably be reborn in a lower realm (because gods squander their time in heaven intoxicated by pleasure), this is thought to be the most intense suffering in the entire cycle of rebirth.

This is the briefest of descriptions of a central topic in Buddhist meditation, one that is often presented in the most gruesome detail. The goal of such meditation is to cause one to regard this life as a prisoner regards his or her prison, to cause one to strive to escape from this world with the urgency that a person whose hair is on fire seeks to douse the flames. The goal of such meditation, in other words, is stress induction. This stress is the result of a profound dissatisfaction with the world. Rather than seeking a sense of peaceful satisfaction with the unfolding of experience, the goal of this practice is to produce a state of mind that is highly judgmental, indeed judging this world to be like a prison. This sense of dissatisfaction is regarded as an essential prerequisite for progress on the Buddhist path. Far from seeking to become somehow "non-judgmental," the meditator is instructed to judge all of the objects of ordinary experience as scarred by three marks: impermanence, suffering, and no self.

With that prerequisite in place, the Buddhist practitioner embarks on a path intended not to reduce stress or lower cholesterol, but to uproot more fundamental forms of suffering. As we have seen, these include what are referred to as the sufferings of pain; in the case of humans, these include birth, aging, sickness, and death, losing friends, gaining enemies, not finding what you want, and finding what you don't want. And the sufferings of pain are only the most overt. The Buddha also spoke of what he called the sufferings of change. These, in fact, are feelings of pleasure, which, by their very nature, will eventually turn into pain. The claim here is that pleasure and pain are fundamentally different: that pain remains painful unless something is done to alleviate it, while pleasure will naturally turn into pain. The most subtle form of suffering of all is one to which the unenlightened are said to be oblivious: that our minds and bodies are so conditioned that we are always subject to suffering in the next moment. This is because our existence depends upon causes and conditions over which we have no control. Karma again.

If the Buddha had sought to alleviate only the most superficial form of suffering, then, endowed with both omniscience and deep compassion for all sentient beings, he would have taught something more compatible with modern science. Rather than teaching something that, to the ears of some, sounds vaguely like the theory of relativity, the Buddha would have provided a cure for suffering as science understands it; two millennia ago he would have set forth the *Indoor Plumbing Sutra* and the *Lotus of Good Dental Hygiene.* But he did not. Instead he spoke of a more subtle suffering, one that for the unenlightened lies unnoticed like a wisp of wool in the palm of the hand, but that for the enlightened can never be ignored, like a wisp of wool in the eye.

Indeed, some Buddhist thinkers have claimed that the teachings of the Buddha are so profound that they will always remain beyond anything that science can understand. The Chinese monk Taixu wrote in 1928, "The reality of the Buddhist doctrine is only to be grasped by those who are in the sphere of supreme and universal perception, in which they can behold the true nature of the Universe, but for this they must have attained the wisdom of Buddha himself, and it is not by the use of science or logic that we can expect to acquire such wisdom."[5] More recently, the famous Sri Lankan monk Walpola Rahula has declared, "to put on the same footing, the Buddha and a philosopher or a scientist, however celebrated he may be, is a gross disrespect to the Great Teacher."[6] Yet for most, the mere fact that scientific research has not yet offered significant insights into Buddhist thought and practice does not mean that it cannot. For the remainder of this chapter, I would like to consider, albeit tentatively, some possible areas for exploration.

Even the most cursory survey of the discourse of Buddhism and science over the past century and a half reveals a series of moments of misrecognition, the philosophical equivalent of approaching a stranger and saying, "You look so familiar. Have we met somewhere before?" Put another way, the history of Buddhism and science is filled with false resonance: the doctrine of karma sounds like the theory of evolution, the Buddhist account of the origin of the cosmos sounds like the Big Bang, the doctrine of emptiness sounds like quantum physics. Immanuel Kant once observed that, "since human reason has been enraptured by innumerable objects in various ways for many centuries, it cannot easily fail that for everything new, something old can be found which has some kind of similarity to it."[7] This is true enough, and it is also true that our minds make consistent use of comparison to organize experience. Comparison may be an evolutionary adaptation. But in the case of Buddhism and science, something else seems also to be at work.

Perhaps like all religions, Buddhism is profoundly retrospec-

tive, looking to the past to understand the present, in the hope
of securing a haven safe from a hazardous future. The Buddha
taught a path from a world of suffering to a state where there is no
suffering. And it is said that he set forth that path only after following it himself. Thus, there is a comfort in the knowledge that
the course to the safe haven has already been clearly charted, there
is a comfort in the knowledge that that course has already been
successfully navigated, there is a comfort in the knowledge that
the other shore has already been reached. Science, in a popular
representation, offers a different appeal, an appeal to the quest
for what has never been known by anyone, and yet is somehow
there, waiting to be discovered, if we just knew how to find it. In
the meantime, we must live in doubt of our deepest knowledge.
Perhaps this is why we yearn for the teachings of the Buddha, an
itinerant mendicant in Iron Age India, to anticipate the formulae
of Einstein. And in this regard, the story of the history of Buddhism and science is different from the history of Christianity and
science. For Christianity, science has often been represented as
a challenge to scripture, especially to the Book of Genesis at the
beginning and the accounts of the resurrection at the end. That
challenge has been powerful, and persistent, and the relation between Christianity and science, at least in its most visible and recent forms, has been characterized as a relation of antagonism.

Why has it been different for Buddhism? It is important to
note that there have been moments of antagonism, although they
are long forgotten. In the first encounters of Buddhism and science in Asia, science was a weapon used by Christian missionaries
against Buddhism, seeking to demonstrate errors in the Buddha's
teaching. During that debate between a Methodist minister and a
Buddhist monk in Sri Lanka in 1873, the Christian speaker asked
why the mountain at the center of the Buddhist universe, said to

be 80,000 *yojanas*—or more than 500,000 miles—high, had never been seen by anyone. According to one account, a Buddhist monk, presumably referring to the Tree of the Knowledge of Good and Evil in the Garden of Eden, shouted from the crowd: "Climb to the top of the tall tree described in your sutras and you will definitely see it."[8] A somewhat more serious response was provided by the Methodist minister's interlocutor, the monk Guṇānanda, who responded with his own question, asking, if the central mountain located to the north of our continent does not exist, then why does the mariner's compass always point north?

But these kinds of disputes over bare facts have occurred only rarely in the history of Buddhism and science. Any number of reasons for this might be put forward, from the storied malleability of Buddhism, so famous for adapting easily to the cultural preconceptions of a given time or place, to the simple fact that Buddhism is far less deeply ingrained in the collective psyche of the West than Christianity. As we saw earlier, Buddhism has often served as a kind of safe surrogate for Christianity, avoiding all those problems of religion and science by being a religion that is also a science. And if it is to serve this purpose, science, whatever that word has meant over the past century and a half, must somehow be found in Buddhism. And so this must sound like that.

This is not to suggest that such parallels do not exist, or that their pursuit is in every case unfruitful. It is rather that before such parallels are proclaimed, something of the long pattern of other parallels must be understood, and something of the psychology of the compulsion to compare must be acknowledged.

Nor is this to suggest in the slightest that research on the neurology of meditation should not be conducted. The prevalence of meditation in Buddhism has long been overstated; the majority of Buddhists over the long history of the tradition have not meditated for a moment. But meditation is the virtuoso practice par excellence of the tradition, and monks have devoted themselves to its practice, and other monks to its theory, for more than two millennia. Clearly something was occurring in their brains, regardless of how it was described, and it would be fascinating to know whether it could be somehow measured. And regardless of how far such neurological data must stray from the traditional instructions for meditation and the meditative states said to result from them, it would be a great loss should the rich vocabulary and imagery of Buddhist meditation somehow be abandoned in the process of scientific research.

There are so many questions to be asked and answered, questions not simply that the ancient meditation practices of Buddhism might answer for modern neurology, but questions about Buddhist meditation that might be answered by neurobiology.

Let me provide one example. Much of what we know about the left brain and the right brain can be traced back to research done in the 1960s by Robert Sperry and William Gazzaniga, working with a patient who, in an effort to stop severe epileptic seizures, had had the *corpus callosum*—the bundle of nerve fibers connecting the right and left sides of his brain—surgically severed, preventing communication between the two sides. The two neurologists designed a series of experiments in which images were displayed to each of the patient's visual fields, asking him to tap keys to indicate what he saw. The right eye and hand are connected to the left brain and the left eye and hand are connected to the right brain. The profound differences between the two hemispheres became clear when images were displayed to the left visual field (and hence the right brain). The patient claimed to have seen nothing, yet his hand (under the control of the right brain) would tap the key each time an image appeared, as he had been instructed to do. It seemed that the right brain had "seen" the image, but had been unable to name it. Subsequent experiments, by these and other scientists, led to the understanding that the right brain is superior to the left brain in dealing with the visual and the spatial, while the left brain is superior to the right brain in dealing with the linguistic and the discursive. As we know, numerous clichés and stereotypes have followed in the wake of this work.

A subsequent experiment with another split-brain patient offered a more nuanced view. The patient was briefly shown two pictures, each visible to one visual field and invisible to the other. The left hemisphere was shown a picture of a rooster's claw; the right hemisphere was shown a house surrounded by snow with a snowman in front. A number of pictures were then placed in front of the patient, and he was asked to select the one that matched the snow scene and the rooster's claw by pointing to the correct picture. For the snow scene, there were pictures of a lawnmower, a rake, a snow shovel, and an ax. For the rooster's claw, there were pictures of an apple, a toaster, a hammer, and a rooster's head. Both hemispheres could see all eight pictures. The left brain, the "logical side," correctly pointed to the picture of the rooster's head to go with the rooster's claw. And the right brain, the "intuitive side," correctly pointed to the picture of the snow shovel to go with the winter scene. In each case, the two hemispheres made the correct association.

However, articulating the reasons for the decision was a different matter. The patient was next asked why the rooster's head and the shovel were selected. The first answer, the left brain answer, was predictable: the rooster's head goes with the rooster's foot. The second answer, the right brain answer, was not. One would expect the patient to explain that you need the snow shovel to shovel snow. But he said, "You need the shovel to clean out the chicken coop." That is, the right brain, the nonverbal brain, knew that the snow shovel goes with the snow, but could not articulate it, could not explain why. The left brain, the verbal brain, never "saw" the snow scene because that visual field was invisible to it; it could only see the claw of the rooster. But the left brain could also see the left hand (which is connected to the right brain) choose the shovel. And so the verbal brain did something creative. It produced a narrative, one that is "logical" under the circumstances, but that is utterly at odds with reality: "You need the shovel to clean out the chicken coop." The left brain seemed compelled to provide the anomalous object with meaning.

There are a number of ways to interpret this story. Perhaps the most cynical of these would be that the stories we tell to explain the world, from the founding myths of the great religions to doctoral dissertations (at least in the humanities), are in the end just so much chicken shit, with all of our efforts at profound analysis just so much shoveling. A somewhat more nuanced reading of the experiment would consider its implications for such perennial questions in the history of philosophy as the role of language in perception and the relation of sense perception and thought, questions that are equally central to Buddhist philosophy. Within the domain of Buddhism and science, the experiment provides a possibly useful perspective on the relation between meditation and doctrine, a question that I would like to briefly explore.

One of the fundamental tropes of Buddhism is that everything the Buddha taught — and he is said to have set forth eighty-four thousand doctrines — is in some sense the enunciation of his silent experience, seated cross-legged in meditation beneath a tree on that moonlit night. The subsequent elaboration of those doctrines, in treatises composed across the Buddhist world — treatises composed in Sanskrit, Pali, Chinese, Japanese, Tibetan, Korean,

Burmese, Thai—is but commentary on the Buddha's words and hence upon his enlightenment. There is thus a profound retrospection in the tradition, as a huge network of texts, in many languages, all trace their roots back to a single, and silent, source. This silence comes into play in the sutras. A text entitled the *Sutra Setting Forth the Inconceivable Secrets of the Tathāgata* (Tathāgata-cintyaguhyanirdeśa) declares, "From the night when the Tathāgata fully awakened into unsurpassed, complete, perfect enlightenment, until the night when he passed without remainder into final nirvana, the Tathāgata did not utter a single syllable."[9]

Another silence descends from another source, as scholars of Buddhism—that is, scholars who used to be called Orientalists—have, for over a century and a half, sought to determine what the Buddha taught. Like Jesus, the Buddha did not write anything; his words were remembered by monks and nuns. Even acknowledging the storied mnemonics of ancient Indian religions, the quest for his true words is complicated by the fact that nothing seems to have been recorded until the last decades before the Common Era, and then, not in India but in Sri Lanka. And the extant versions of what is accepted by the tradition as *buddhavacana*, the speech of the Buddha, derive from Sri Lankan editions from the fifth century C.E., at least eight centuries after the Buddha's death. And none of those manuscripts survive; the oldest extant manuscripts of the Pali canon (regarded by many as the best record of the Buddha's teachings) date from the early ninth century. The oldest Buddhist manuscripts, some of which date from the first century B.C.E., were recently discovered in Afghanistan. Some scholars continue to think that the Buddha's original teachings can be reconstructed; others consider his words to be unrecoverable. Both the Asian tradition and the Western scholarship look back to a certain silence.

This silence has led to a focus on—some would say a fetish-
ization of—experience. In the field of religious studies there has
been considerable debate on the relative priority of mystical experience and doctrine, on whether doctrine is somehow the articulation of mystical experience, or whether that experience is
both induced and defined by doctrine. William James, as one
might expect, gave priority to experience, writing in *The Varieties
of Religious Experience:* "The truth is that in the metaphysical and
religious sphere, articulate reasons are cogent for us only when
our inarticulate feelings of reality have already been impressed
in favor of the same conclusion. Then, indeed, our intuitions and
our reason work together, and great world-ruling systems, like
that of the Buddhist or of the Christian philosophy, may grow
up. Our impulsive belief is here always what sets up the original
body of truth, and our articulately verbalized philosophy is but
its shadowy translation into formulas. The unreasoned and immediate assurance is the deep thing in us, the reasoned argument
is but the surface exhibition."[10]

In the discourse of Buddhism and science, there has been almost universal support for the priority of experience, stemming in
part from the Victorian view of the Buddha as a man who gained
enlightenment through his own silent reflections, rather than
through divine revelation that was conveyed verbally, as it was to
Moses and Muhammad. The Buddha's mind is thus portrayed as
a laboratory where hypotheses are tested and rejected, arriving
finally at the truth, which the Buddha then articulated. Leaving
aside the question of how the Buddha's brain might be different
from ours, what might the rooster's claw have to do with the question of the relation between doctrine and experience, in this case,
the experience of meditation?

If it is the case, as research has suggested, that only about

2 percent of the information that the brain takes in each minute is processed consciously, with the other 98 percent processed by the unconscious brain, what are the mechanisms that allow the unconscious to become conscious? What are the mechanisms that allow the conscious to penetrate and illuminate the unconscious? In the case of meditation, how does silent experience become verbalized? These are profound questions in the history of philosophy, made all the more difficult by problems of translation, where terms like "conscious" and "unconscious" have no immediate correlates in the extensive Buddhist vocabulary of mental states. If these questions are considered from the perspective of the rooster's claw, what confidence do we have that the narrative deriving from the experience (whatever that term might mean) of meditation accurately reflects that experience? If doctrine derives from experience, then the experiment suggests that doctrine is, at best, epiphenomenal. In the course of neurological research, one would expect that the deepest states of Buddhist experience will eventually be identified as the firing of neurons in a specific area of the brain. What is the relation of this experience to the classical description of it found in the Buddhist canon? The massive edifice of Buddhist doctrine would seem in some sense an afterthought, an inadequate and in some sense irrelevant attempt to articulate what the brain has already understood at a nonverbal level, an understanding that cannot be spoken.

In order to consider this question, one would need to identify whether meditation is a right brain (that is, "intuitive") or left brain (that is, "rational") activity. Any survey of the standard meditation manuals would suggest that it is both. For example, there is a traditional division into analytical meditation and serenity meditation. As described earlier, analytical meditation would include reasoned reflection on the absence of a self among the physical and mental constituents of the person. Serenity meditation would include the various techniques for quieting thought and focusing one-pointedly on a given object.

How might these categories be mapped onto the neurobiological categories of analysis and insight? Here, analysis is defined as a methodical search for a solution to a problem; it is a process located in the left brain. Insight is defined as breaking through an impasse, with affects of both surprise and certainty; insight has been located in the right brain. Experiments have shown that such insight is inhibited both by verbalization and by concentration. Insights occur at unexpected moments, especially during states of drowsiness after waking from sleep. When anyone interested in Buddhist meditation hears the word "insight," she or he thinks immediately of the Pali term *vipassanā*, translated as "insight," and the name of one of the most popular forms of Buddhist meditation in the West. Didn't the Buddha become enlightened after meditating all night? He might have been drowsy. And unexpected moments? That sounds like Zen.

But again, caution is called for. In Buddhist literature, insight is not exactly spontaneous and its knowledge is not exactly unexpected. Unlike arriving at a new scientific discovery—one thinks of the famous story of August Kekulé's vision of the Ouroboros, a snake biting its own tail, when he discovered the benzene ring—Buddhist insight involves arriving at an understanding that many others have gained in the past, and, furthermore, an understanding that one has already arrived at intellectually.

Yet, it must also be said that much Buddhist literature is suspicious of the intellect, seeing it as a tool to be used and then abandoned. Describing the use of analysis in the cultivation of wisdom, the *Kāśyapa Chapter Sutra* (Kāśyapaparivarta) says: "Kāśyapa, it is thus. For example, fire arises when the wind rubs two branches together. Once the fire has started, the two branches are burned. Just so, Kāśyapa, if you possess the correct analytical intellect, noble wisdom is created. Through its creation, the analytical intellect is consumed." And it is said that the Buddha has no thought, only direct perception without the medium of mental images. So it seems as if the goal is the realization of the silent intuition associated with the right brain. And yet the Buddha speaks.

Like the Buddha's original teachings, the neurology of the Buddha's enlightenment is irretrievable. And so the question of the origin of doctrine remains in the domain of myth. But Buddhist monks and nuns have meditated for millennia, and they have done so based on discursive instructions, a discourse that has long claimed to result in the deepest states of awareness of which the human mind is capable. If there is to be a dialogue between Buddhism and science, it will come in the form of translation, an activity central to the spread of the dharma over the centuries and across the realms. Here, the translation will not be the translation of the four levels of the formless realm into levels of blood pressure, but the more challenging translation of doctrine into meditative states and meditative states into scientific data.

What, then, can we say about the Scientific Buddha? First, he is a young buddha, born in the nineteenth century, not in the royal city of Kapilavastu, but in the republican city of Paris. From there he traveled to Sri Lanka and then to other parts of Asia. A century and a half after his birth, he is alive and well in the imaginations of many, all around the world. And yet, from a Buddhist perspective, he is out of place.

According to Buddhist doctrine, there have been many buddhas in the past, and there will be many buddhas in the future. The Buddha who appeared in India in the fifth century B.C.E., our Buddha, is the fourth or the seventh or the twenty-fifth. There are long intervals between one buddha and another. Indeed, it is said that all traces of a previous buddha—his followers, his relics, his teachings—must have disappeared from the world before a new buddha appears. This is because there is no reason for a new buddha to come while the teachings of the past buddha continue to benefit the world. It is only when the teachings of the last buddha, and even his name, have been completely forgotten that a new buddha appears, reminding the world of the buddhas who have come in the past, predicting the buddha who will come in the future, and teaching the same path to the same nirvana that all buddhas have taught over time without beginning.

Time in Buddhism is not cyclic, as is often claimed. Worlds come in and out of existence, in phases of creation, abiding, destruction, and nothingness. Beings wander among the six realms. Yet time moves forward to a time when there is no time, when *saṃsāra* itself comes to an end. Despite the confusion that seems to surround us, there is movement forward.

This cosmic order is disrupted by the Scientific Buddha. He appeared in the world before the teachings of the Buddha of our age, Śākyamuni Buddha, had been forgotten, before his teachings had run their course. The Scientific Buddha was not predicted by a previous buddha, nor did the world await his coming. And yet he has served a useful role. He was born into a world of the colonial subjugation of Asia by Europe. He fought valiantly to win Buddhism its place among the great religions of the world, so that today it is universally respected for its values of reason and nonviolence. We might regard the Scientific Buddha as one of the many "emanation bodies" of the Buddha who have appeared in the world, making use of skillful methods (*upāya*) to teach a provisional dharma to those temporarily incapable of understanding the true teaching. For this buddha was stripped of his many magical elements and his dharma was deracinated. The meditation that he taught was only something called "mindfulness," and even then, a pale form of that practice. That is, he taught something that no other buddha in the past had taught: stress reduction.

Some scholars have begun to refer to a form of Buddhism called "modern Buddhism." Its origins are hazy, but most would agree that it has existed since at least the mid–nineteenth century. Despite now having endured for almost two centuries, not an insignificant fraction of the entire history of Buddhism, modern Buddhism has been seen as something of a monolith, with its own defining doctrines, one of which is the compatibility of Buddhism and science. But it may be the case that modern Buddhism has existed long enough to require its own periodization, one in which each period of modern Buddhism has its own favored "science." Such an approach would transform what I have described as a problem—that the compatible science keeps changing—into a key element of the modern history of Buddhism, or at least of the history of modern Buddhism. Cultures change and religions change over time, often with great rapidity. Buddhism has also changed, with the persistent claim in the tradition—a tradition that has spoken so eloquently of impermanence—that the teachings have remained the same providing perhaps the most compelling proof of such change. This is a historical fact. But the question here is a different one: What becomes of the old buddha when a new buddha appears on the scene, a buddha whose advent was unheralded? What becomes of the teachings of that old buddha, the pre-scientific buddha?

Previous buddhas had increased stress, explaining, "Monks, all is burning," in the *Fire Sermon;* that we are trapped in a house on fire, in the *Lotus Sutra;* that we should regard the world as a prisoner regards his prison on the night before his execution. Previous buddhas sought to create stress, to destroy complacency, in order to lead us to a state of eternal stress reduction, that state of extinction called nirvana.

The Scientific Buddha is a pale reflection of the Buddha born in Asia, a buddha who entered our world in order to destroy it. This buddha has no interest in being compatible with science. The relation of Buddhism and science, then, should not be seen as a disagreement over when and how the universe began. It should not be seen, in Stephen Jay Gould's memorable phrase, as "non-overlapping magisteria," with science concerned with fact and religion concerned with morality. It should not be seen, in Buddhist terms, as the two truths, with science concerned with the conventional truth, and Buddhism concerned with the ultimate truth. Buddhism and science each have their own narrative, each their own *telos*. If an ancient religion like Buddhism has anything to offer science, it is not in the facile confirmation of its findings. If the past has a future, it is in its description of an alternative world, one that calls into question so many of the fundamental assumptions of our scientific world. A Tibetan poet wrote these lines about the Buddha:

> To the sharp weapons of the demons, you offered delicate flowers
> in return.
> When the enraged Devadatta pushed down a boulder, you practiced
> silence.
> Son of the Śākyas, incapable of casting even an angry glance at your
> enemy,
> What intelligent person would honor you as protector from fearful
> saṃsāra?[11]

"To the sharp weapons of the demons, you offered delicate flowers in return." This line refers to the story of the Buddha's enlightenment, when Māra, the god of desire and death, dispatched his army of demons against the Buddha, attacking him with a hail of weapons. The Buddha transformed their hail of spears and arrows into a gentle rain of flowers. "When the enraged Devadatta pushed down a boulder, you practiced silence." This refers to the Buddha's evil cousin, Devadatta, who, seeking to succeed the Buddha as head of the order of monks and nuns, tried to assassinate the Buddha by pushing down a boulder to crush him. The boulder only nicked the Buddha's toe, making a small cut. By his act, Devadatta committed two of the deeds that result in rebirth in hell: causing blood to flow from the Buddha and causing dissent in the monastic community. But in neither attack did the Buddha retaliate. "Son of the Śākyas, incapable of casting even an angry glance at your enemy, what intelligent person would honor you as protector from fearful saṃsāra?" A Buddhist is someone who seeks refuge from suffering in the three jewels, the Buddha, the dharma, that is, the Buddha's teachings, and the *saṃgha*, the community of the enlightened. Here the poet asks, if this is how the Buddha responded to attacks on his own person, what intelligent person—the Tibetan term is *blo ldan*, perhaps better translated as "what person in his or her right mind"—would look to the Buddha for protection from the myriad sufferings of *saṃsāra*, the Buddha who would not even defend himself?

The poet's suggestion is that the purpose of the Buddhist path is to overturn the ordinary; the poet's implication is that being in one's right mind has been the problem all along. Thus, the Buddha does the opposite of what the wise would counsel, setting aside his own welfare in lifetime after lifetime, constantly resisting the demands of self, declaring in fact that there is no self, compassionately leading all beings to their own extinction. And as the result of his resistance to what the world regards as wisdom, the Buddha achieved the ultimate state of perfection.

The Buddha did the opposite, and he also taught the opposite. From his perspective, the perspective of being awake (the literal meaning of *buddha* in Sanskrit), the world is asleep. From his perspective, what the world believes to be true is in fact false, what the world believes to be right side up is upside down. And so he identified what he called "the four upside down views." They are pleasure, permanence, purity, and self. That is, what the world perceives as pleasurable is ultimately painful. What the world perceives as permanent is in fact transitory. What the world perceives as pure is in fact impure. Where the world perceives a self, there is no self.

The Buddha said that humans are subject to eight forms of suffering: birth, aging, sickness, death, losing friends, making enemies, not finding what you want, and finding what you don't want. How is it possible that a mendicant living in India in the Iron Age could have said something that was also true more than two millennia later, in the Information Age? This is the question that confronts us so often in study of the past, a past from which we are abstracted by time, by space, by language. There is so much about the past that seems incommensurable with our own time. Those things that seem familiar often seem so only because of our own projections, projections that themselves derive from a long process of history. And yet, through the miracle of translation, we still weep at the death of Enkidu in the *Epic of Gilgamesh*, we are stirred by Odysseus's slaughter of the suitors, we laugh out loud at the antics of a magic monkey in China in *Journey to the West*. There is something that has remained the same, something that has remained true, as long as we do not look too hard to find where that truth might reside. There is something mysterious about it, like trying to translate the brow of a whale. Melville's chapter concludes with these words: "Champollion deciphered the wrinkled granite hieroglyphics. But there is no Champollion to decipher the Egypt of every man's and every being's face. Physiognomy, like every other human science, is but a passing fable. If then, Sir William Jones, who read in thirty languages, could not read the simplest peasant's face, in its profounder and more subtle meanings, how may unlettered Ishmael hope to read the awful Chaldee of the Sperm Whale's brow? I but put that brow before you. Read if it you can."

The Buddha does not need to be preserved in aspic, all of his wondrous aspects kept intact, frozen in time, the founder of a dead religion. At the same time, the Buddha does not need to be brought up to date, his teachings do not need to be reinterpreted into terms utterly foreign to what he taught, or what his tradition says he taught. Something is always lost in translation, simply in the rendering of a word from one language to another. In order to limit that loss, we might let what the tradition says about the Buddha be heard, allowing him to keep his forty teeth, his hair that needs no cutting, the bump on his head. Rather than imagining things about the Buddha and his teachings for which there is no evidence, we might dwell on what is there, and the ways in which these things might somehow continue to bear meaning. The preservation of the mythological and the miraculous is not merely a matter of aesthetics. At least two questions, pondered by Buddhists over the centuries, remain worthy of our contemplation: "What does it mean to seek the welfare of others?" and "Is there a self?"

One of the most famous statements in Buddhist literature occurs in the *Diamond Sutra*, where the Buddha says to the monk

Subhūti, "In this regard, Subhūti, one who has set out on the bodhisattva path should have the following thought, 'I should bring all living beings to final extinction in the realm of extinction without substrate remaining. But after I have brought living beings to final extinction in this way, no living being whatsoever has been brought to extinction.' Why is that? If, Subhūti, the idea of a living being were to occur to a bodhisattva, or the idea of a soul or the idea of a person, he should not be called a bodhisattva. Why is that? There is no *dharma* called 'one who has set out on the bodhisattva path.'"

That is, the Buddha says that the bodhisattva is someone who vows to liberate all beings in the universe from suffering, while understanding that ultimately there are no beings in the universe to be liberated from suffering because those beings do not ultimately exist; they, like the bodhisattva, have no self. This is obviously a bold statement. It is the claim that there can be compassion without essentialism. It is here, perhaps, in the domain of wisdom and compassion—the essential elements of the path to buddhahood, the two wings of the bird that flies to enlightenment—that the incompatibility of Buddhism and science becomes most clear.

In the field of developmental psychology, Susan Gelman has recently published important research on essentialism. She has shown that from a very young age, children in a wide range of cultures come to believe that the objects that constitute a particular category share an essence, a true nature, something that is not visible but that nonetheless provides each member of that category with its own identity as well as its shared membership in the category.[12]

For any student of Buddhism, this immediately evokes the notion of *svabhāva*, a Sanskrit term literally meaning "own being" and variously translated as "intrinsic nature," "inherent existence," even "essence." It is the claim of the Madhyamaka school of Buddhist philosophy that all beings—whether animals, humans, or gods—instinctively believe that all phenomena, all objects of their experience, both inanimate and animate, are endowed with some intrinsic nature that makes each object what it is. This perception of an essence is not only conceptual, an idea, but extends even to nonverbal sense experience; we see things, hear things, smell things, taste things, and touch things as if they had their own essence.

Gelman seeks to understand why children see the world in this way, concluding that it is an evolutionary adaptation, one that has certain benefits for interacting with the objects of our experience. For the Buddhist, of course, the conception of intrinsic existence, or essentialism, is the most fundamental form of ignorance, the most basic of all errors, and hence the root cause of all suffering. In this sense, the belief in essences is also evolutionary because it is the fuel that powers the engine of birth, death, and rebirth. It is this belief in essences that must be destroyed in order to bring an end to suffering and rebirth. Wisdom, in Buddhism, is the understanding that there is no essence, that there is no self.

The Buddhist sutras tell of a hell called Hell of the Fiery Chariot, where the damned are made to serve as beasts of burden for eons, dragging a fiery chariot driven by a demon. Long ago, two denizens of hell were pulling a chariot driven by a huge demon with fire shooting from his eyes, with smoke pouring from his mouth, nose, and ears. He beat them with an iron club. One of the tormented beings was unable to go any further. Enraged, the demon attacked him with a trident. He screamed in pain, crying out to his father and mother. The other denizen of hell felt compassion for his companion and stood in front of the demon, asking him to stop. The demon, becoming more enraged, stabbed him in the neck and killed him. It is said that that single moment of compassion for another person, a compassion felt although he himself was suffering the same torments of hell, cut short his long lifetime in hell that otherwise would have lasted for eight eons. Billions of years later, that denizen of hell became the Buddha.[13]

Anyone destined to buddhahood—and according to some important Mahayana sutras, all beings are destined to buddhahood—must develop compassion. Research on the psychopathic personality suggests that empathy is genetic. Yet, the Buddhist claim is that empathy, and compassion, can be developed by anyone, and eventually will be developed by everyone. The Buddhist traditions do not assume that compassion is a natural endowment, a personality trait possessed by some and lacked by others. Compassion is a state of mind that can, and must, be cultivated. Far from seeking to quiet habitual thoughts, the bodhisattva is obsessed by one. It is the thought, "I will liberate all beings in the universe from suffering." It must be the first thought in the morning and the last thought at night; it must be the thought that motivates all actions. And techniques are provided for producing this habitual thought.

One of the most renowned of these is set forth in a long poem entitled *Entering the Path to Enlightenment* (Bodhicaryāvatāra), by the eighth-century Indian monk Śāntideva. The technique is called "the exchange of self and other." He explains, "Whatever happiness there is in the world all arises from the wish for others' happiness. Whatever suffering there is in the world all arises from the wish for one's own happiness."[14] From one perspective, this is simple karma theory. Concern for one's happiness can lead to all manner of negative deeds, whether they be mental, like covetousness; verbal, like lying; or physical, like stealing. These are deeds that will produce suffering in the future. Concern for the happiness of others can lead to all manner of virtuous deeds, whether they be mental, like wishing others well; verbal, like speaking kindly; or physical, like charity. These deeds will produce happiness in the future. But Śāntideva seems to regard the exchange of self and other as something more than a simple code of ethics. He writes, "Whoever wishes to protect oneself and others quickly should practice the most sublime of secrets: the exchange of self and other."[15] Śāntideva then goes on to describe what the exchange of self and other means. In short, it means to cherish others as one once cherished oneself, and to ignore oneself as one once abandoned others.

His claim is that this is the secret technique for the achievement of buddhahood. His proof is the Buddha himself. All other beings in the universe have devoted all of their energies to staying alive, to finding happiness, to securing their own future. Each being has striven for its own survival, seeking happiness but finding only suffering in yet another of the myriad forms of existence within *saṃsāra*, for eternity. Long ago, the Buddha was such a being. Yet at some point, perhaps in that hell, he made the decision to stop seeking his own welfare and sought the welfare of others. Today, he is liberated from the cycle of birth and death and has found the state beyond all suffering. As Śāntideva says, "What need is there to say more? The childish act for their own welfare. The Sage acts for the welfare of others. Behold the difference between them."[16]

Biologists have studied what appears to be altruism in animals. The standard example of such behavior would be an animal sounding the alarm at the approach of a predator, thus protecting the herd but also calling the predator's attention to itself and thus threatening its own survival. In seeking to explain this behavior, scientists have appealed to game theory, in which mutually beneficial behavior serves the perpetuation of the species. The common example is the grooming that birds and mammals perform, removing parasites from places on the body that an individual is unable to reach. Among the various possibilities of a particular interaction between two animals, there would be a number of scenarios, the most common being reciprocation, with each animal grooming the other in turn. There might also be cases, however, in which one animal does not reciprocate. In the biological literature on the subject, the animal who receives grooming but does not groom is referred to as a cheat. The animal who grooms but does not receive grooming is referred to as a sucker. According to evolutionary theory, the cheat has the better chance of survival because he has lessened his chances of infection by a disease-carrying parasite. His genes will eventually spread through the population, while the suckers die off. As a result, the population will eventually evolve into a population of cheats.

Śāntideva's implication is that a different process of extinction begins at the moment when one person feels compassion for another. This moment of compassion eventually results in the achievement of buddhahood, and from that state one leads all others to extinction. The sucker is exalted to the rank of savior.

The appeal that we continue to remember the Buddha in the various ways that he has been understood over the long history of Buddhism in Asia is not to suggest that Mount Meru can be found using GPS or that Noah's Ark will ever be unearthed. This is not to claim that Buddhist descriptions of the world carry the same status as the descriptions of the most current scientific research (that is, those descriptions that have not yet been displaced). Nor is it to consign the Buddha to some vague realm of "the ultimate," conceding all else to "the conventional." It is to say, instead, that the Buddha, the old Buddha, not the Scientific Buddha, presented a radical challenge to the way we see the world, both the world that was seen two millennia ago, and the world that is seen today. What he taught is not different, it is not an alternative, it is the opposite. That the path that we think will lead us to happiness leads instead to sorrow. That what we believe is true is instead false. That what we imagine to be real is unreal. A certain value lies in remembering that challenge from time to time.

To understand oneself, and the world, as merely a process, an extraordinary process of cause and effect, operating without an essence, yet seeing the salvation of others, who also do not exist, as the highest form of human endeavor. This is the challenge presented by that passage from the *Diamond Sutra*. The scientific verification of this bold claim would seem to lie, like buddhahood itself, far in the future.

Notes

CHAPTER ONE: A PURIFIED RELIGION

1. See Stanley L. Jaki, ed., *Lord Gifford and His Lectures: A Centenary Retrospect* (Edinburgh: Scottish Academic Press, 1995), pp. 99–100.

2. Paul Carus, "Buddhism and the Religion of Science," *The Open Court* 10 (March 12, 1896): 4845.

CHAPTER TWO: THE BIRTH OF THE SCIENTIFIC BUDDHA

1. *The Book of Ser Marco Polo the Venetian Concerning the Kingdoms and Marvels of the East*, 2 vols., trans. and ed. by Sir Henry Yule, 3rd ed., revised by Henri Cordier (New York: AMS, 1986), reprint of 1926 London edition, vol. 2, pp. 316–317. See also the extensive notes of Yule and Cordier, pp. 320–330.

2. Athanasius Kircher, *China Illustrata*, trans. by Charles D. Van Tuyl (Bloomington, Ind.: Indiana University Research Institute, 1987), pp. 141–142.

3. Engelbert Kaempfer, *The History of Japan: Together with a Description of the Kingdom of Siam, 1690–1692*, vol. 1 (London: J. MacLehose and Sons, 1906), pp. 62–66.

4. As late as 1810, Edward Moor would write: "Some statues of BUDDHA certainly exhibit thick *Ethiopian* lips; but all, with wooly hair: there is something mysterious, and still unexplained, connected with

the hair of this, and only of this, *Indian* deity. The fact of so many different tales having been invented to account for his crisped woolly head, is alone sufficient to excite suspicion that there is something to conceal — something to be ashamed of; more exists than meets the eye." Edward Moor (1771–1848), *The Hindu Pantheon*, facsimile of original 1810 London edition (New York: Garland Publishing, 1984), pp. 231–232.

5. "Sur quelques épithètes descriptives de Bouddha qui font voir que Bouddha n'appartenait pas a la race nègre," in Jean-Pierre Abel-Rémusat, *Mélanges Asiatiques*, vol. 1 (Paris: Librarie Orientale de Dondey-Dupré Père et Fils, 1825), pp. 100–128.

6. William Erskine, "Account of the Cave-Temple of Elephanta with a Plan of the Drawings of the Principal Figures," *Transactions of the Literary Society of Bombay*, vol. 1 (1819): 201–202.

7. Eugène Burnouf, *Introduction to the History of Indian Buddhism*, trans. by Katia Buffetrille and Donald S. Lopez, Jr. (Chicago: The University of Chicago Press, 2009), pp. 180–181.

8. Léon Feer, *Papiers d'Eugène Burnouf conservés à la Bibliothèque Nationale* (Paris: H. Champion, 1899), pp. 158–159.

9. Raymond Schwab, *The Oriental Renaissance: Europe's Rediscovery of India and the East, 1680–1880* (New York: Columbia University Press, 1984), p. 439.

10. Burnouf, *Introduction to the History of Indian Buddhism*, p. 328.

11. *The Middle Length Discourses of the Buddha: A New Translation of the Majjhima Nikāya*, original translation by Bhikkhu Ñāṇamoli; translation edited and revised by Bhikkhu Bodhi (Boston: Wisdom Publications, 1995), p. 167.

CHAPTER THREE: THE PROBLEM WITH KARMA

1. Thomas W. Rhys Davids, *Lectures on the Origin and Growth of Religion as Illustrated by Some Points in the History of Indian Buddhism*, Hibbert Lectures, 1881 (London: Williams and Norgate, 1881), p. 94. Claims of the compatibility of Buddhism and evolution were common in the Victorian and Edwardian periods. We read in a 1905 essay in *The Fortnightly Review*, "The religion of the Buddha is not in conflict with modern science; he anticipated many of its most important conclusions; its primary prin-

ciple of evolution is one with his central tenet." W. S. Lilly, "The Message of Buddhism to the Western World," *The Fortnightly Review* (new series) 78: (July to December, 1905): 213.

2. See *The Literary Digest*, May 31, 1890, p. 22 (162).

3. Thomas H. Huxley, *Evolution and Ethics and Other Essays* (London: Macmillan and Company, 1894), p. 61.

4. Walter Y. Evans-Wentz, *The Fairy-Faith in Celtic Countries*, p. 515, n. 1. Learning that according to standard Buddhist doctrine, rebirth is not a process of ascent through stages of rebirth but an apparently random wandering among realms of various degrees of divinity and damnation, Evans-Wentz, in works like *The Tibetan Book of the Dead*, would explain that this was merely the exoteric teaching intended for the uninitiated. Initiates of esoteric Buddhism (which was another name for Theosophy) knew that regression through the rounds of rebirth was impossible. For a study of this text, see Donald S. Lopez, Jr., *The Tibetan Book of the Dead: A Biography* (Princeton: Princeton University Press, 2011).

5. See the *Cūḷakammavibhaṅga Sutta* in the *Middle Length Discourse of the Buddha*, trans. by Bhikkhu Ñāṇamoli and Bhikku Bodhi (Boston: Wisdom Publications, 1995), p. 1053.

6. Joseph Dalton Hooker, *Himalayan Journals; or, Notes of a Naturalist in Bengal, the Sikkim and the Nepal Himalayas, the Khasia Mountains, &c*, vol. 1 (London: John Murray, 1854), p. 303.

7. Charles Darwin, *The Life and Letters of Charles Darwin Including an Autobiographical Chapter*, ed. by Francis Darwin, vol. 1 (New York: D. Appleton and Company, 1897), p. 278.

8. Ibid., pp. 279–280.

9. Ibid., p. 280.

10. Ibid., p. 279.

11. His Holiness the Dalai Lama, *The Universe in a Single Atom: The Convergence of Science and Spirituality* (New York: Morgan Road Books, 2005), p. 112.

12. Ibid., p. 131.

13. J. Robert Oppenheimer, *Science and the Common Understanding* (New York: Simon and Schuster, 1954), p. 43.

14. Nāgārjuna and the Kaysang Gyatso, Seventh Dalai Lama, *The*

Precious Garland and The Song of the Four Mindfulnesses (New York: Harper and Row, 1975), p. 45.

15. Étienne Lamotte, *History of Indian Buddhism from the Origins to the Śaka Era*, trans. by Sara Webb-Boin (Louvain: Institut Orientaliste, 1998), p. xxiv.

16. Patrick Olivelle, trans., *Life of the Buddha by Ashva-ghosha*, Clay Sanskrit Library (New York: New York University Press, 2008), pp. 137, 139.

INTERLUDE: A PRIMER ON BUDDHIST MEDITATION

1. D. J. Gogerly, "On Buddhism," *Journal of the Ceylon Branch of the Royal Asiatic Society* 1, no. 1 (1845): 24.

2. R. Spence Hardy, *A Manual of Budhism in Its Modern Development* (London: Partridge and Oakey, 1853), p. 412.

3. Thomas W. Rhys Davids, trans., *Buddhist Suttas* (Oxford: The Clarendon Press, 1881), p. 107.

4. Thomas W. Rhys Davids, trans., *The Questions of King Milinda*, vol. 1 (Oxford: The Clarendon Press, 1890), p. 58.

5. Nyanaponika Thera, *The Heart of Buddhist Meditation: A Handbook of Mental Training Based on the Buddha's Way of Mindfulness, with an Anthology of Relevant Texts translated from the Pali and Sanskrit* (San Francisco: Weiser Books, 1965), pp. 7–9.

CHAPTER FOUR: THE DEATH OF THE SCIENTIFIC BUDDHA

1. Samuel Landgon, *Quarterly Letters Addressed to the General Secretaries of the Wesleyan Methodist Missionary Society* 82 (September 1873): 198. The passage is cited in R. F. Young and G. P. V. Somaratna, *Vain Debates: The Buddhist-Christian Controversies of Nineteenth-Century Ceylon* (Vienna: Publications of the De Nobili Research Library, 1996), p. 175, n. 381.

2. Henry S. Olcott, *The Buddhist Catechism*, 44th edition (Adyar, India: The Theosophical Publishing House, 1947), p. 91.

3. Maria B. Ospina, et al., "Meditation Practices for Health: State of the Research," Evidence Report/Technology Assessment No. 155. AHRQ Publication No. 07-E010 (Rockville, Md.: Agency for Healthcare Re-

search and Quality, June 2007): 6. Even a recent paper claiming significant results contains the following caveat in its penultimate paragraph:
"It should be noted also that MBSR [Mindfulness Based Stress Reduc-
tion] is a multifaceted group program and some positive effects may result from components not specific to meditation or mindfulness, such as
group social interaction, stress education, or gentle stretching exercises."
Britta K. Hözel, et al., "Mindfulness Practice Leads to Increases in Re-
gional Brain Gray Matter Density," *Psychiatry Research: Neuroimaging* 191
(2011): 42.

4. See Donald S. Lopez, Jr., ed., *Religions of Tibet in Practice*
(Princeton: Princeton University Press, 1997), pp. 433–434.

5. His Eminence Tai Hsu, *Lectures in Buddhism* (Paris, 1928), pp. 47–
48.

6. Ven. Dr. Walpola Rahula, "Religion and Science," *Dharma Vijaya:
Triannual Publication of Dharma Vijaya Buddhist Vihara Los Angeles* 2, no. 1
(May 1989): 14.

7. Cited in Wilhelm Halbfass, *India and Europe: An Essay in Understanding* (Albany: State University of New York Press, 1988), p. 112.

8. The story is recounted in Bkras mthong thub bstan chos dar, *Dge
'dun chos 'phel gyi lo rgyus* (Dharamsala: Library of Tibetan Works and
Archives, 1980), p. 59.

9. For the Sanskrit see Louis de la Vallée Poussin, ed., *Mūlamadhya-
makakārikās de Nāgārjuna avec le commentaire Prasannapadā de Candrakīrti*
(Osnabruck: Biblio Verlag, 1970), p. 366.

10. William James, *The Varieties of Religious Experience* (New York:
Collier Books, 1961), pp. 74–75.

11. See Donald S. Lopez, Jr., *The Madman's Middle Way: Reflections
on Reality of the Tibetan Monk Gendun Chopel* (Chicago: The University of
Chicago Press, 2006), p. 47.

12. Susan A. Gelman, *The Essential Child: Origins of Essentialism in
Everyday Thought* (Oxford: Oxford University Press, 2003).

13. See *Thabs mkhas pa chen po sangs rgyas drin lan bsab pa'i mdo* (Toh.
353). In the Derge edition of the Tibetan canon; the passage is found in
Mdo sde, vol. 76 (A:), 117a4–117b5.

14. *Bodhicaryāvatāra* VIII.129. This and the following translations

are made from the Tibetan. For an English translation of the entire text, see Śāntideva, *The Bodhicaryāvatāra*, trans. by Kate Crosby and Andrew Skilton (Oxford: Oxford University Press, 1996).

    15. *Bodhicaryāvatāra*, VIII.120.

    16. *Bodhicaryāvatāra*, VIII.130.

Index

Abel-Rémusat, Jean-Pierre, 30
"Access concentration," 85, 88
Afghanistan, 116
Agency for Healthcare Research
and Quality, 105-106
Altruism, 130-131
American Oriental Society, 39-40
Analytical meditation, 87-88,
118-119
Aristotle, 102
Aryan languages, 41
Asia: Buddhism in, 16, 34; and
colonialism, 11-12; representa-
tions of the Buddha, 24, 26-27,
29, 30
Astrology, 103
Aśvaghoṣa, 80

Babur (Mughal emperor of India),
31
Bhāvanā, 83
Blavatsky, Helen Petrovna, 49-50

Bodhisattvas, 91, 127, 129
Brain, 14, 19, 113-115, 118
British East India Company, 30-
31, 34, 36
Brooks, David, 9
Buddha: ability to create illusions,
104; African hypothesis, 29, 30,
133-134n4; birth of, 26, 27, 29,
34, 40; death of, 7, 91; enlightenment of, 13, 14-15, 43, 45,
57-58, 76, 79, 82, 93, 116, 117,
119, 120, 123-124; European encounters with, 21-25; European
quest for historical Buddha,
16-17, 39-40; four noble truths
taught by, 5, 41, 57-62, 66-67,
89; four upside down views,
125; identity of, 15, 16-17; investigation and analysis of,
6-7, 14; meditative posture of,
82; mind of, 43; and multiple
universes, 6-7; and nature of

Buddha (continued)
reality, 75–76, 87; and no self, 6, 67, 77, 124, 125, 127; operations of mind, 6; Pali biographies of, 42; path revealed by, 44, 57–58, 60, 67–68, 70, 72, 75, 76, 79, 80, 87, 111, 124, 131–132; prescience of, 13; qualities of, 42–43; renunciation of royal birth, 41, 80; smile of, 42–43, 77; speech of, 43, 116; translation of name, 24–27; and unindicated views, 70; validation of, 45. *See also* Śākyamuni Buddha; Scientific Buddha

Buddhas: birth of, 56–57; and Buddhist doctrine, 120–121; early literature on, 45; in Mahayana doctrine, 91; and nature of reality, 75–76; and path to enlightenment, 76; preceding Śākyamuni Buddha, 43–44, 45; and stress, 122–123

Buddhavacana, 116

Buddhism: academic study of, 38; authentic forms of, 12; canonical languages of, 33; and causation, 5–6; changing meaning of, 13–16, 18, 79, 104, 122; Christianity compared with, 11, 81, 112; Christian missionaries' attacks on, 10–11, 27, 40, 75, 111–112; debate with Christianity, 103; and epistemology, 75–76; European conceptions of, 11, 16, 23–24, 25, 27,

29–30, 31, 33, 34; identified with mindfulness, 14, 79; innovation in, 44–45; and mind/matter dualism, 69–71, 77; modern Buddhism, 122; monastic tradition in, 68, 76, 81–82, 84, 86, 97–98, 113, 116, 120; as purified religion, 5, 8; tradition of the marvelous in, 78. *See also* Tibetan Buddhism; Zen Buddhism

Buddhism/science discourses: and Blavatsky, 49–50; and Buddha's analysis, 6–7, 14; Buddhism as purified religion, 5, 8; and Buddhist doctrine, 5, 85, 109–110; and Buddhist path, 131–132; and choice between religion and science, 19; and Christian missionaries' attacks on Buddhism, 10–11, 40, 111–112; claims for compatibility, 9–11, 13–14, 46, 48–49; and cosmic order, 56–57; and Dalai Lama, 12, 14, 17, 51; and evolution, 49, 76–78, 79, 110, 134–135n1; and identity of Buddha, 15, 16–17; and meaning of Buddhism, 15–16, 18, 104; and meaning of science, 13, 14, 18; and meditation research, 18–19, 104–105; and modern Buddhism, 122; priority of experience, 117; pseudoscience linked to Buddhism, 104; and scientific Buddha, 33; Sri Lanka debate on, 103; and

Tibetan Buddhism, 12, 104; and translation, 120, 126

Buddhist doctrine: and buddhas, 120–121; and Buddha's silence, 116–117, 118; and Buddhism/science discourses, 5, 85, 109–110; and Buddhist meditation, 86, 115–116, 119–120; four noble truths, 5, 41, 57–62, 66–67, 89, 96, 97; and happiness, 66–67; and natural selection, 17, 80; and nirvana, 60, 62, 121; purpose of, 14; reality of, 109–110; and rebirth, 69–70, 74–75, 77, 135n4; and split-brain research, 19; and suffering, 5, 44, 58–62, 66–68, 73, 109, 111. *See also* Karma

Buddhist meditation: analytical meditation, 87–88, 118–120; and breath, 93–94, 96, 98; and Buddha's enlightenment, 79; and Buddhist doctrine, 86, 115–116, 119–120; and Buddhist path, 60, 82, 87, 90–91, 94, 95, 108; Buddhist texts on, 84; and concentration, 60, 83, 85–87, 88, 89, 90, 92, 93, 95; and enlightenment, 85–86; forms of, 83–84, 106; in history of Buddhism, 81–84; insight as aim of, 84, 87–89, 92, 95, 98, 119; laypeople and, 81, 97–98, 112–113; Mahayana embellishments to, 91; meanings of, 91–92, 105; and mindfulness, 92–97, 98; and

monastic tradition, 81–82, 84, 86, 97–98, 113, 120; and neurobiology, 18–19; and no self, 87, 88, 89, 90, 96, 108; and operations of mind, 6; path of accumulation, 88–89; path of meditation, 90; path of preparation, 89; path of vision, 89–90; premises of, 84; and rebirth, 85, 86–87, 90, 107, 108; serenity as aim of, 84–85, 87, 88, 118, 119; and *smṛti*, 93–96, 98; and suffering, 84, 88–89, 96, 108–109; and three trainings, 95; and translation of Sanskrit terms, 82–83; as universal practice, 99. *See also* Meditation

Buddhist poetry, 107, 123–124, 129–130

Buddhist texts: and Buddha's qualities, 42–43; on karma, 52–53; on meditation, 84; and philology, 38–39, 41

Burma: Buddhist meditation in, 97–99; representations of the Buddha, 24

Burnouf, Eugène, 36–38, 39

Campbell, Archibald, 63

Capra, Fritjof, 9

Carus, Paul, 7

Ceylon. *See* Sri Lanka

Chambers, Robert, 48

Charles XI (king of Sweden), 27

China: Buddhism in, 12, 34, 40; Buddhist meditation in, 83, 92;

China (continued)
  and Kublai Khan, 22; representations of the Buddha, 24, 25, 26, 28
Chinese Communist Party, 12
Christianity: Buddhism compared with, 11, 81, 112; debate with Buddhism, 103; and science, 10–11, 103, 111, 112; secular scholarship in theology, 3
Christian missionaries, attacks on Buddhism, 10–11, 27, 40, 75, 111–112
Clement of Alexandria, 21, 22
Colonialism, 10, 11–12, 121
Compassion, 64, 129, 131
Concentration, 60, 83, 85–87, 88, 89, 90, 92, 93, 95, 119
Consciousness: and Buddhist path, 80; continuum of, 72; and cosmic order, 56; Infinite Consciousness, 86; and mind/matter dualism, 69, 70–71
Creationism, 17

Dahui, 92
Dalai Lama (fourteenth): and Buddhism/science discourses, 12, 14, 17, 51; and evolution, 17, 50, 69; and mind/matter dualism, 69
Darwin, Charles: and compassion, 64; and evolution, 17, 47, 50–51, 68–69; and happiness, 65, 66, 67; and Hooker, 63; and natural selection, 65–66, 80; and reli-

gion, 64–65; and suffering, 18, 61, 65, 66, 67–68
Desideri, Ippolito, 24
Dharma, 7, 56, 77, 96–97, 107, 120, 121–122, 127
Dhyāna, 83
Diamond Sutra, 126–127, 132
Discourse on the Foundations of Mindfulness, 95–96, 98
Duff, Alexander, 2
Duff Lectures, 2
Dutch East India Company, 28, 30–31

Egypt, 29–30
Einstein, Albert, 7–8, 9, 15, 111
Ekman, Paul, 50–51, 64
Emerson, Ralph Waldo, 38
Entering the Path to Enlightenment (Śāntideva), 129–130, 131
Epicurus, 31, 33, 37
Epistemologies, 75–76
Erskine, William, 31–35, 36
Essentialism, 127–128
Ethics: ethical deeds, 60, 72, 129–130; and three trainings, 60, 95; as virtue, 77. *See also* Morality
Evans-Wentz, Walter, 50, 135n4
Evolution: and Buddhism/science discourses, 49, 76–78, 79, 110, 134–135n1; and extinction, 72–73, 75, 77; and karma, 17–18, 49, 68–69, 76–77, 110; and rebirth, 49, 71–72; and religion, 47; social evolution, 11. *See also* Natural selection

The Expression of the Emotions in Man and Animals (Darwin), 50, 64

Extinction: and Buddhist path, 73–74, 75, 79, 80, 127, 131; and natural selection, 72–73, 75, 77

The Fairy-Faith in Celtic Countries (Evans-Wentz), 50

Formless Realm, 85, 86–87, 89, 90

Fowler, Lorenzo, 103

Fowler, Orson, 103

Freud, Sigmund, 106

Gall, Franz Joseph, 102, 103

Gama, Vasco da, 34

Game theory, 130–131

Garland of Jewels (Nāgārjuna), 72

Gazzaniga, William, 113

Gelman, Susan, 127, 128

Gifford, Adam, Lord, 2, 3

Gifford Lectures, 2

Gogerly, Daniel J., 93–94

Gould, Stephen Jay, 123

Great Britain: India as colony of, 31, 39; and phrenology, 102–103; and Sri Lankan independence, 12

Great Discourse on the Lion's Roar (Mahāsīhanāda Sutta), 45

Guṇānanda, 112

Hardy, Robert Spence, 10, 94

Haskell, Caroline, 2

Haskell Lectures, 2

The Heart of Buddhist Meditation (Nyanaponika Thera), 99

Hells: and Buddha's smile, 42; and Buddhist cosmology, 35, 77; and Buddhist doctrine, 46; and Buddhist meditation, 107–108; Hell of the Fiery Chariot, 128–129; and karma, 53, 54; and rebirth, 58–59, 72

Hibbert, Robert, 2

Hibbert Lectures, 2, 49

Himalayan Journals (Hooker), 63

Hinduism, 30, 31, 34, 38, 44, 64, 93

Hodgson, Brian Houghton, 36, 37, 63

Hooker, Joseph Dalton, 51, 62–63

Huxley, Thomas, 49

Idols and idolaters, in European conception of Buddhism, 11, 16, 23–24, 25, 27, 29–30, 31, 33, 34

Ignorance: and conception of self, 59, 73–74; and karma, 57; as motivation for action, 6; and suffering, 58, 59, 60, 67

India: as British colony, 31, 39; and Buddhism, 16, 33, 34, 39; Buddhist meditation in, 82, 88, 90–91, 92, 99; and European encounter with Buddha, 21–22, 24; representations of the Buddha, 26, 27, 30–32, 33, 34, 40, 134n4; tantric practices in, 92

Indo-European languages, 41

Infinite Consciousness, 86

Infinite Space, 86

"In Memoriam" (Tennyson), 47–48

Insight meditation, 83, 84, 87–89, 92, 95, 98

Introduction à l'histoire du Buddhisme indien (Burnouf), 38

Jacquet, Eugène, 36

Jainism, 31

James, William, 117

Japan: Buddhism in, 12, 34, 40; and Christian missionaries, 11, 40; Pure Land tradition of, 92; representations of the Buddha, 24, 26, 27, 28–29, 33; and Zen Buddhism, 13

Jerome, Saint, 21–22

Jesus, historical, 16

John of Damascus, Saint, 22

Jones, William, 29, 126

Kaempfer, Engelbert, 27–29, 30, 31–32, 33, 36

Kant, Immanuel, 110

Karma: and Buddhist meditation, 107–108, 109; classical doctrine of, 17–18, 51–57, 59, 67; complete and incomplete actions, 53–55; and cosmic order, 55–56; and evolution, 17–18, 49, 68–69, 76–77, 110; and extinction, 73; and intention, 53–54; and moment of death, 54–55; natural law of, 6, 33, 55–56,

129; and natural selection, 51, 68–69, 71; physical universe as product of, 55, 75; and rebirth, 53, 54, 55, 57, 68–69, 72, 74

Kāśyapa Chapter Sutra, 119–120

Kekulé, August, 119

Kircher, Athanasius, 25–27, 29, 36

Knox, Robert, 24

Koans, 92

Korea, 11, 34

Kublai Khan, 22, 23

Lamaism, 12, 104. *See also* Tibetan Buddhism

Lamotte, Étienne, 78

Lang, Andrew, 2

Langdon, Samuel, 103

Lavater, Johann Kaspar, 102

Ledi Sayadaw, 98

Life and Letters of Charles Darwin (Darwin), 64–65

Literary Society of Bombay, 31

Lotus Sutra, 37, 122

Loubère, Simon de la, 24

Louis XIV (king of France), 24

Lubac, Henri de, 40–41

Mackintosh, James, 31

Madhyamaka school, 128

Mahayana school, 91, 129

Maitreya Buddha, 76

Majjhima Nikāya, 51

Maṇḍala, 92

Mantras, 83

Materialist Buddhism, 18

Meditation: common view of,

90; definition of, 105; modernization of, 81; as right brain or left brain activity, 118–120; and Scientific Buddha, 122; scientific research on, 18–19, 81, 92, 99, 104–106, 112–113, 118–120; and stress, 18, 97, 106, 108, 122; therapeutic effects of, 9, 97, 105–106. *See also* Buddhist meditation

Meiji Japan, and Buddhism, 12

Melville, Herman, 101, 103, 126

Memory: in Buddhism, 79, 93. *See also* Mindfulness

Mengden, Nicolai Alexandrovitch, 65, 67

Merchants' Lectures, 1

Mindfulness: Buddhism identified with, 14, 79; and Buddhist meditation, 92–97, 98; four establishments of, 88, 95–97, 98; and Scientific Buddha, 122

Mindfulness Based Stress Reduction (MBSR), 106, 137n3

Mind/matter dualism, and Buddhism, 69–71, 77

Moby-Dick (Melville), 101, 103, 126

Möngke Khan, 22

Mongols and Mongolia, 22, 34

Monier-Williams, Monier, 94

Moor, Edward, 133–134n4

Morality: and Buddha's teachings, 17, 35, 38, 41; and causation, 5–6; and karma, 55–56, 59. *See also* Ethics

Müller, Friedrich Max, 2, 94

Nāgārjuna, 72, 104

Nārada, U, 98

National Center for Complementary and Alternative Medicine, 105–106

Natural selection: and Buddhist doctrine, 17, 80; and Darwin, 65–66, 80; and karma, 51, 68–69, 71; and perpetuation of species, 71; and survival of species, 73, 74

Neither Existence nor Nonexistence, 86

Neuroscience, 18–19, 78, 104–105, 112–115, 118–119

Newtonian physics, 13

Nietzsche, Friedrich, 38

Nirvana: Buddha's attainment of, 86; and Buddhist doctrine, 60, 62, 121; and Buddhist meditation, 97, 106; and Buddhist path, 57, 72, 73, 77–78; Freud on, 106; and monastic tradition, 68

Niyamadīpanī, 55

No self: and Buddha, 6, 67, 77, 124, 125, 127; and Buddhist meditation, 87, 88, 89, 90, 96, 108; and wisdom, 60, 87, 128

Nothingness, 86

Nyanaponika Thera, 98–99

Olcott, Henry Steel, 50, 104

Open Court Press, 7

Oppenheimer, J. Robert, 70

Pali canon, 13, 42, 45, 55, 104, 116

Perfection of Wisdom sutras, 91

Philology, 38–39, 41

Phrenology, 102–103

Polo, Marco, 22–23, 27, 33–34

Popper, Karl, 103

Pratītyasamutpāda (dependent origination), 6

Pseudoscience, 103, 104

Pure Land tradition, 92

Pythagoras, 26, 37, 102

Quantum mechanics, 13

Race, and theories of language groups, 41

Rahula, Walpola, 110

Realm of Desire, 85, 86, 88, 89, 90

Realm of Form, 85, 86, 87, 89, 90

Rebirth: beginningless cycle of, 71; Blavatsky on, 50; Buddhist doctrine on, 69–70, 74–75, 77, 135n4; and Buddhist meditation, 85, 86–87, 90, 107, 108; and Buddhist path, 57–58, 60, 67–68, 80, 87; classical Buddhist proof of, 69; and evolution, 49, 71–72; and hells, 58–59, 72; and karma, 53, 54, 55, 57, 68–69, 72, 74; states of, 58–59

Reciprocation, 131

Reincarnation, 26, 30

Religion: changes in, 78–79, 122; Einstein on, 7–8, 9; and science, 3, 4, 7–8, 10, 17, 19, 47. *See also* Buddhism; Christianity

Rhys Davids, Thomas W., 49, 94

Ricci, Matteo, 24, 25

Royal Asiatic Society, 29, 94

Ruggieri, Michele, 25

Sacred Books of the East, 94

Śākyamuni Buddha: buddhas preceding, 43–44, 45; and Marco Polo's account, 23; and Scientific Buddha, 121; and translations of Buddha's name, 24

Salisbury, Edward Eldridge, 39–40

Samādhi, 83, 91, 95

Saṃgha, 124

Saṃsāra, 57, 74, 85, 106, 107–108, 121, 124, 130

Sanskrit language: and Buddhist meditation, 82–83; Burnouf's study of, 36–38; and philology, 41; Salisbury's study of, 39–40; Whitney's study of, 40

Śāntideva, 129–130, 131

Satipaṭṭhana Sutta, 95–96, 98

Schelling, Friedrich, 38

Schopenhauer, Arthur, 38

Science: and altruism, 130–131; changing meaning of, 13, 14, 18; and Christianity, 10–11, 103, 111, 112; and epistemology, 75; and physiognomy, 101–102, 126; and religion, 3, 4, 7–8, 10, 17, 19, 47. *See also* Buddhism/science discourses

Scientific Buddha: birth of, 21, 25, 35, 39; and Burnouf, 36–38, 40; identity of, 17, 120; role of, 121–122; and Salisbury, 39–40; story of, 44; and traditional Buddha, 45–46, 121, 123; transformation of Buddha from god to man, 33, 38, 40; validation of, 45

The Secret Doctrine (Blavatsky), 50

Serenity meditation, 6, 84–85, 87, 88, 92, 118, 119

Siddhārtha, Prince, 57–58, 79–80

Smṛti, 93–94, 95. *See also* Mindfulness

Social evolution, 11

Socinians, 1

Socrates, 37

Southeast Asia, 13, 34, 104

Sozzini, Fausto, 1

Sperry, Robert, 113

Spiritualism, 49

Split-brain research, 19, 113–115

Spurzheim, Johann Gaspar, 102, 103

Sri Lanka (Ceylon): Buddhism in, 13, 22–23, 34, 40, 104; and Buddhist meditation, 82, 99; and Christian missionaries, 11, 40; debate between Christianity and Buddhism, 103; independence from Great Britain, 12; representations of the Buddha, 23, 24; and Scientific Buddha, 120

Stress: and buddhas, 122–123; and meditation, 18, 97, 106, 108, 122

Subhūti, 127

Śuddhodana, King, 80

Suffering: and bodhisattvas, 127; Buddhist doctrine on, 5, 44, 58–62, 66–68, 73, 109, 111; and Buddhist meditation, 84, 88–89, 96, 108–109; of change, 61–62; of conditioning, 62; Darwin and, 18, 61, 65, 66, 67–68; forms of, 61, 109, 125; four aspects of, 89; and karma, 6, 52, 57; of pain, 61

Śūnyatā (emptiness), 6

Sutra Setting Forth the Inconceivable Secrets of the Tathāgata, 116

Suzuki, D. T., 104

Svabhāva, 127–128

Taixu, 109–110

Tang Dynasty, 24

The Tao of Physics (Capra), 9

Taylor, Jill Bolte, 9

Tennyson, Alfred, Lord, 47–48

Terry, Dwight, 1, 2–5, 7, 8, 17, 47

Terry Lectures, 1, 2–4

Thailand, representations of the Buddha, 24, 28–29, 30, 32

Theosophical Society, 49–50

Theravada Buddhism, 13, 104

Thomas Aquinas, Saint, 23

Thoreau, Henry David, 38

Tibet: Buddhism in, 34; independence of, 12; representations of the Buddha, 24, 34–35

The Tibetan Book of the Dead (Evans-Wentz), 50, 135n4

Tibetan Buddhism: Buddhist meditation in, 106-107; and Darwin, 50-51, 64; and Hodgson, 63; and Hooker, 63; and poetry, 123-124; and science, 12, 13-14, 104; texts of, 63-64

Transmigration of souls, 26, 30, 35

The Universe in a Single Atom (Dalai Lama), 69

U.S. Department of Health and Human Services, 105

Vedas, 44

Vestiges of the Natural History of Creation (Chambers), 48

Vietnam, 24, 27

Wagner, Richard, 38

Warburton Lectures, 1-2

Warren, Henry Clarke, 70

Whitney, William Dwight, 40

Wisdom: analysis in cultivation of, 119-120; of Buddha, 110, 124-125; and Buddhism/science discourses, 127; and destruction of ignorance, 74; and insight meditation, 87; and karma, 57; nobility derived from, 41; and no self, 60, 87, 128; object of, 91; state of perfect wisdom, 7; and three trainings, 60, 95; as virtue, 77

Xavier, Francis, 10, 24, 25

Yuan Dynasty, 22

Zen Buddhism: and Japan, 13; and meditation, 82, 119; and Suzuki, 104